Peptides

Biology and Chemistry

Peptides

Biology and Chemistry

Proceedings of the 1998 Chinese Peptide Symposium
July 14-17, 1998, Lanzhou, China

Edited by

Xiao-Yu Hu
Department of Biochemistry and Molecular Biology,
School of Life Science,
Lanzhou University,
Lanzhou, China

Rui Wang
Department of Biochemistry and Molecular Biology,
School of Life Science,
Lanzhou University,
Lanzhou, China

and

James P. Tam
Department of Microbiology and Immunology,
Vanderbilt University,
Nashville, TN, U.S.A.

Springer-Science+Business Media, B.V.

A C.I.P. Catalogue record for this book is available from the Library of Congress.

ISBN 978-94-017-3875-0 ISBN 978-0-306-46880-3 (eBook)
DOI 10.1007/978-0-306-46880-3

Printed on acid-free paper

Preface

The Fifth Chinese Peptide Symposium, hosted by Lanzhou University, was held at Lanzhou, China July 14-17, 1998, with 156 participants, including 30 scientists from abroad, representing nine countries. The four-day conference was both intense and spiritually rewarding. Our goal for CPS-98 was to provide a forum for the exchange of knowledge, cooperation and friendship between the international and Chinese scientific communities, and we believe this goal was met.

The symposium consisted of 8 sessions with 42 oral and 90 poster presentations, including synthetic methods, molecular diversity and peptide libraries, structure and conformation of peptides and proteins, bioactive peptides, peptide immunology, De Novo design and synthesis of proteins and peptides, ligand-receptor interactions, the chemistry-biology-interface and challenging problems in peptides.

The enthusiastic cooperation and excellent contributions were gratifying and the active response of the invited speakers contributed to the success of the symposium. The presentations were of excellent caliber and represented the most current and significant aspects of peptide science.

Dr. Kit Lam of the University of Arizona and Dr. Yun-Hua Ye of Peking University were the recipients of "The Cathay Award" sponsored by the H. H. Liu Education Foundation, offered for their seminal contribution in peptide science and the Chinese Peptide Symposium. Four outstanding young scientists were selected by the organizing committee to receive awards sponsored by Haikou Nanhai Pharmaceutical Industry Co. Ltd. (Zhong He Group).

It is our pleasure to acknowledge Nobel Laureate Professor Bruce R. Merrifield for his kindness in supporting and serving as the chairman of the Awarding Committee for the "Cathay Award," and all members of the program and organizing committee for their generous contributions to the successful symposium.

Kluwer Academic Publishers kindly agreed to publish the CPS-98 proceedings, and we thank them for their contributions and support of the Symposium. We greatly appreciate the generous financial assistance of the sponsors and donors, in particular the major sponsors from abroad, The American Peptide Society and Eli Lilly and Company. On behalf of the organizing committee, we thank the administrative assistance by Lanzhou University, as well as Vanderbilt University.

Xiao-Yu Hu
Rui Wang
James P. Tam

Chinese Peptide Symposium – 1998

July 14-17, 1998, Lanzhou, China
Lanzhou University

Chairpersons

Prof. Xiao-Yu Hu, *Lanzhou University, Lanzhou, China*
Prof. Rui Wang, *Lanzhou University, Lanzhou, China*

Program Committee

Chairpersons

Xiao-Yu Hu, *Lanzhou University, China*
Rui Wang, *Lanzhou University, China*
James Tam, *Vanderbilt University, U.S.A.*

Members

Subaro Aimoto, *Institute for Protein Res., Osaka University, Japan*
Meng-Shen Cai, *Beijing Medical University, China*
Yu-Cang Du, *Shanghai Institute of Biochemistry, China*
Victor Hruby, *University of Arizona, U.S.A.*
Yoshiaki Kiso, *Kyoto Pharmaceutical University, Japan*
Kit Lam, *Arizona Cancer Center, U.S.A.*
Gui-Shen Lu, *Insitute of Materia Medica, Beijing, China*
Jean Martinez, *CNRS Faculte de Pharmacie, France*
Arnold Satterthwait, *The Scripps Research Insitute, U.S.A.*
John Wade, *University of Melbourne, Australia*
Jie-Cheng Xu, *Shanghai Insitute of Organic Chemistry, China*
Xiao-Jie Xu, *Peking University, China*
Yun-Hua Ye, *Peking University, China*
You-Shang Zhang, *Shanghai Institute of Biochemistry, China*

Awarding Committee of Cathay Award

Prof. Bruce Merrifield, Chairman, *The Rockefeller University, U.S.A.*
Prof. Victor J. Hruby, *University of Arizona, U.S.A.*
Prof. James Tam, *Vanderbilt University, U.S.A.*
Prof. Yu-Cang Du, *Shanghai Institute of Biochemistry, China*
Prof. Gui-Shen Lu, *Institute of Materia Medica, Beijing, China*
Prof. Xiao-Jie Xu, *Peking University, China*
Prof. Xiao-Yu Hu, *Lanzhou University, China*

Major Sponsors

National Natural Science Foundation of China
School of Life Science, Lanzhou University
K. C. WONG Education Foundation, Hong Kong
American Peptide Society
Eli Lilly & Company

Sponsors

Ministry of Education of China
The Association of Science and Technology of Gansu Province of China
Perkin-Elmer Applied Biosystems Division
American Peptide Company, Inc.
Penninsula Laboratories, Inc.
C. S. Bio. Co.
Zhonghe Pharmaceutical Company
Chengdu Di-ao Pharmaceutical Company
H.H. Liu Education Foudation

Donors

Biomeasure Incorporated
Phoenix Pharmaceuticals, Inc.
ASTA Medica AG
UCP Belgium
Chinese Society of Biochemistry & Molecular Biology

Contents

Session B: Structure, folding and conformation analysis/ De novo design of peptide & protein

Session C: Neuro/Endocrino/Bioactive peptide

Abbreviations

α-MSH α-Melanotropin stimulating hormone
Acm acetamidomethyl
AMMC(But)MIVE
ARDS adult respiratory distress syndrome
ATEE acetyl tyrosine ethyl ester
BK brady Kinin
BOC *tert*-butyloxycarbonyl
BOP (benzntriazol-1-yloxy)-tris(dimethylamino) phoshponium hesafluorophosphate
BTC bis(trichloromethyl) carbonate
Bu butyl
CCK-B cholecystokinin-B
CE capillary electrophoresis
CHO Chinese Hamster Ovary
CoMFA comparative molecular field analysis
CRF corticotropin releasing-factor
DAP-aa N-(o,o-dialkyl) phosphoxylacted amino acids
DBU 1,8-diazabichyclo[5.4.0.]-undec-7-ene
DCM dichloromethane
DEPBT 3-(diethoxyphosphoryloxy) – 1,2,3-benzotriazin-4(3H)-one
DIEA *N,N*-diisopropylethylsulfonic acid
DIPP diisopropyl phosphoryl
DISCO distance-comparisons
DMF dimethylformamide
DMSO dimethyl sulfoxide
DPDPE c[Dpen2, Dpen5]enkephalin
DPP dipropylphosphoryl
DTI Destetrapeptide insulin
ECT Ecl calcitonin
EDT 1,2-ethanedithiol
ELISA enzyme-linked immunoadsorbent assay
Endo-M endo-h-N-acetylglucosaninidase of Mucor hiemalis
ES-MS electric spray mass spectrometry
FAB fast atom bombardment mass spectrometry
FAB-MS
FAB-MS
Fmoc fluoren-9-ylmethoxycarbonyl
FPTase farnesyl protein transferase
FSH follide stimulating hormone
GH growth hormone
GHRP growth hormone releasing peptide
GLCNAC glycopeptide containing the Asn
GMP140 α-granule membrane protein
GPCR G-protein coupled receptor
GPCR G-protein coupled receptor
GST glutathione transferases
hCT human calcitonin
Hep heptyl
Hib-PS Haemophilus influenzae b polysaccharide
HNMR protein nuclear magnetic resonance
Hobt c3-hydroxy-3,4-dihydro-4-oxo-1,2,3-benzotriazine
HOOBt *N*-hydroxyoxodihydrobenzotriazine
HPLC high performance liquid chromatography
HRMS high-resolution negative ion FAB mass spectrum
IPTG isopropyl-D-thiogalactoside
Lys-BK Kallidin
MALCTOF-MS matrix-assisted laser desorption mass spectrometry
MALDI-TOF MALDI time-of-
MAS magic angle spinning

MBHA methylbenzhydrylamine
Mbzl 4-methylbenzyl
MCR melanocotin receptor
MDP N-acetyl muramy –L-alanyl –D-isoglutamine
MMT-1 mouse metallothionein class I
Mpt-MA dimethylphosphinothioic mixed anhydride
MS flight mass spectrometry
MT metallothionein
Mut methanol-utilization slow
MVD mouse vas deferens
NMP N-methylprolidinone
NMR nuclear magnetic resonance
NOS nitric oxide synthase
OGP Osteogenic Growth Peptide
PCR polymerase chain reactions
PEG polyethylene glycol
Pen β,β-dimethylcysteine
Ph Phenol
PhAcOZ p-phenylacetoxybenzyloxycarbonyl
PIP porcine insulin precursor
PLA2 phospholipase A2
PLC phospholipase C
PNA peptide nucleic acid
POEPOP polyoxyethylene-polyoxypropylene
POMC proopiomelanocortin gene
Pq cells transformed by PICQ without PIP gene
QSAR quantitive structrure-activity relationship
RIA radioimmunoassay
RP-HPLC reverse-phase high performance liquid chromatography
SCLC small cell lung carcinoma
sCT salmon calcitonin
sCT salmon calcitonin
SDS-PAGE sodium dodecyl sulfate PAGE
TFA trifluoroacetic acid
TFMSA trifluoromethylsulfonic acid
THF tetrahydrofuran
THP triple-helical peptide
TLC thin layer chromatography
TMT β-methyl-2',6'-dimethyltyrosine
TTT tetrahydrothiazole-2-thione

Session A
Novel approach on synthetic method

Chairs: Yoshiaki Kiso
Kyoto Pharmaceutical University
Yamashina-ku, Kyoto, Japan

Rui Wang
Lanzhou University
Lanzhou, China

Orthogonal ligation of free peptides

James P. Tam, Yi-An Lu and Qitao Yu
Department of Microbiology and Immunology, Vanderbilt University, Nashville, TN 37232, USA

Over past eight years, our laboratory has focused on developing new methodologies for ligating free peptide segments to form complex proteins and biopolymers in aqueous or organic solutions. Recently, we and others have developed a novel segment ligation strategy [1-3] in which an amide bond is formed regiospecifically to the desired N-terminal amine between unprotected peptide segments containing more than one free N-terminal amine. This strategy represents a significant methodological advance. We refer it as "orthogonal ligation strategy" in accordance with other orthogonal concepts in chemistry, including orthogonal protection schemes [4], orthogonal activation [5] and coupling [6] in organic chemistry that distinguish two functional sites based on chemoselectivity.

The orthogonal ligation is a cascade consisting of two reactions of capture and activation. The capture step utilizes the principle of chemoselective ligation to form a covalent intermediate between two peptide segments. Then, an amide bond is formed via an intramolecular acyl transfer through entropic activation (fig. 1). The intramolecular acylation rate, which is first order and often spontaneous, minimizes side reactions associated with enthalpic activation methods.

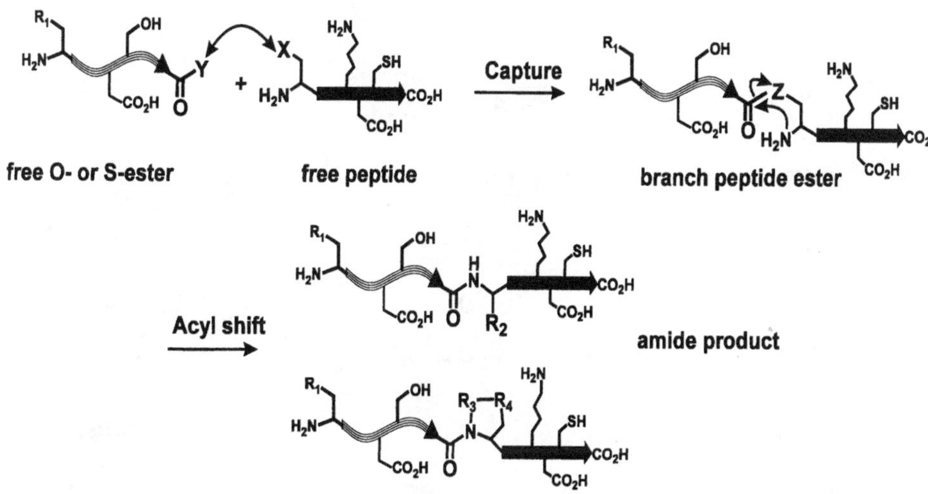

Fig. 1. General concept of orthogonal ligation. X and Y represent a nucleophile and electrophile pair in the capture step to give Z, which provides a side chain R_2 after an acyl shift reaction.

The conceptual framework of orthogonal ligation is similar to the "Prior thiol capture" introduced by Kemp and his coworkers in their "Prior thiol capture" strategy [7-9]. However, there are three fundamental differences between them. 1) It does not require a template to bridge the two peptide segments and is not restricted by a thiol capture mechanism to form a covalent tricyclic bridged intermediate; 2) Instead of a >12-member ring, the acyl migration is mediated through a more favorable five- or six-member ring intermediate; 3) Finally, it fully exploits unprotected peptide segments as building blocks. These improvements provide greater versatility and practicality to segment ligations.

All orthogonal ligation methods have an intramolecular acyl transfer reaction, but their capture methods vary. They fall into two general categories. In the imine capture, the initial capture product is an imine intermediate formed between the amine and acyl segments that contain an aldehyde electrophile. Imine ligation involves a capture step that results from an acyl-aldehyde with an N-terminal amine on an amine- segment (fig. 2). This imine then undergoes a rapid ring-chain tautomerization due to the addition of a nucleophile paired to the N-terminal amine. The resulting heterocycle facilitates an acyl transfer of the ester intermediate to form a proline-like imidic bond. The effectiveness of this strategy has been confirmed in the synthesis of analogs of 50-residue TGF_{α} [10,11], 99-residue peptide HIV-1 protease [12], and 59-residue antimicrobial peptide bactenecin 7 [13].

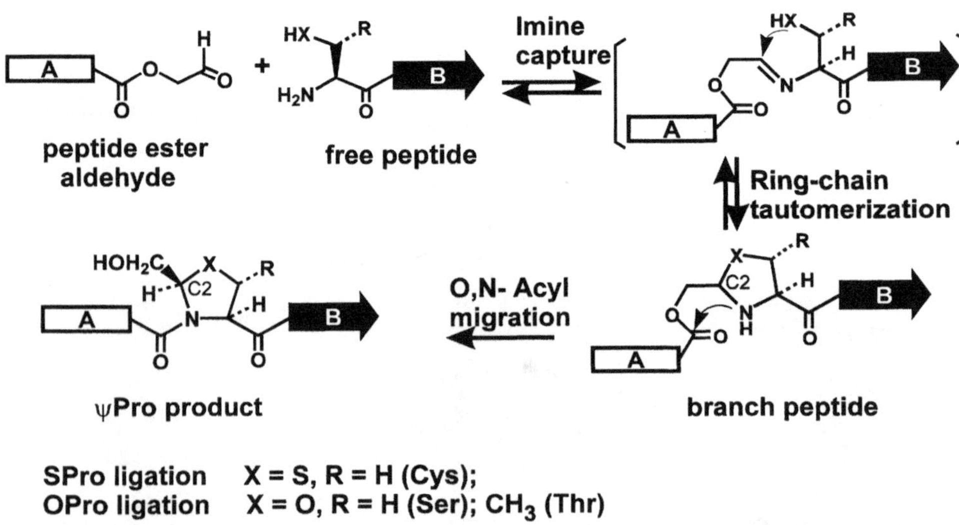

Fig. 2. Imine ligation to form pseudoprolines.

In the thioester capture ligation, a C-terminal peptide thioester readily reacts with a thiol through transesterification to form a new thioester. With an N-terminal Cys, this thioester linkage forming an S-acyl covalent intermediate spontaneously undergoes an S to N-acyl migration to form a peptide bond through a five-member ring intermediate (fig. 3). As a method of capture in orthogonal ligation, the thioester method has been applied successfully for cysteine [2, 3], methionine [14], glycine [15] and histidine ligation of unprotected peptides. Among them, the Cys-thioester ligation, which regenerates a Cys at the ligation site, has been most extensively exploited in chemical and, more recently, semi-syntheses of proteins [16-18].

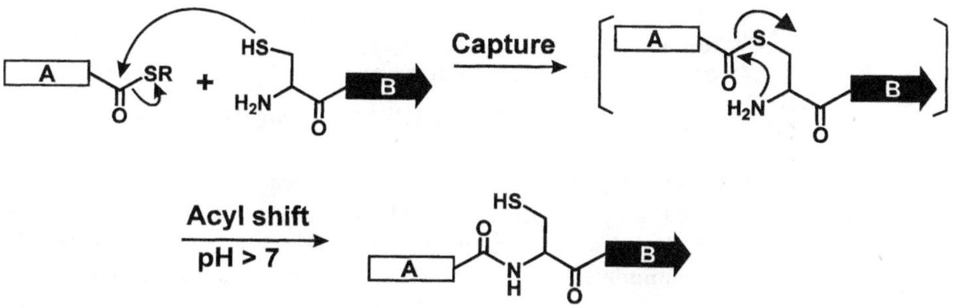

Fig. 3. Thioester ligation.

The principle of orthogonal ligation has been used to develop to a new orthogonal cyclization strategy that forms an amide bond intramolecularly. In orthogonal cyclization, a pair of mutually reactive functionalities are placed at either the N- and C-terminal or at a side-chain and one terminus of a linear unprotected peptide precursor. After capture and acyl migration steps, end-to-end or end-to-side chain cyclic peptide products are formed (fig. 4).

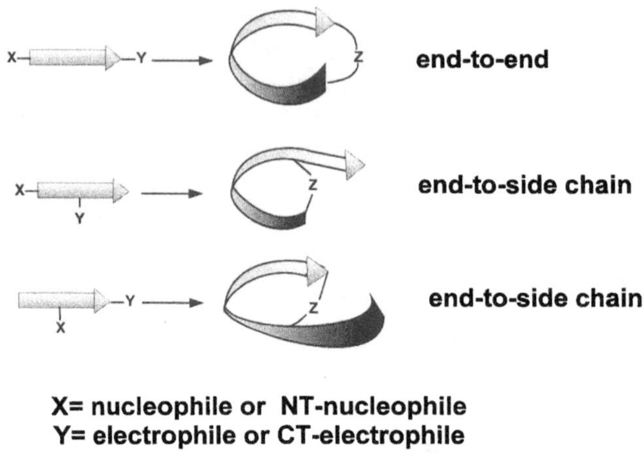

Fig. 4. Concepts in orthogonal cyclization to form amide bond linkage in the ligation site.

Similar to intermolecular ligation, intramolecular imine capture with an N-terminal Cys and C-terminal glycoaldehyde followed by O, N-acyl migration leads to a cyclic lactam as a final product. A series of peptides derived from the third variable (V3) loop on gp120 of human immunodeficiency virus (HIV-1) has been synthesized by this method [19].

Again similar to thioester ligation, cysteine cyclization of an N-terminal Cys and a C-terminal thioester is mediated by creation of an intramolecular bond to form a thiolactone. The NT-thiolactone is a very active intermediate and spontaneously forms the lactam through a five-member ring via an S, N-acyl migration. For the synthesis of large cyclic peptides or proteins with multiple disulfide bonds, thia zip cyclization through the thioester ligation offers a facile approach [20]. The thia zip reaction proceeds through a series of small thiolactone intermediates by successive ring expansions at the carbonyl ring junctions analogous to the aza-zip reaction. These reversible equilibrations of thiol-thiolactone exchanges ultimately result in the formation of an N-□terminal amino thiolactone linking the N and C termini. This large N-terminal amino thiolactone intermediate readily undergoes a ring contraction through an S to N acyl transfer that gives an end-to-end lactam.

Thus far, more than a dozen orthogonal ligation methods based on either imine or thioester captures have been developed to afford native and unusual amino acids at ligation sites of linear, branched or cyclic peptides. Because unprotected peptides and proteins of different sizes and forms can be obtained from either chemical or recombinant sources, orthogonal ligation removes the size limitation imposed on the chemical synthesis of a protein with a native or non-native structure. Furthermore, by using building blocks from biosynthetic sources, orthogonal ligation provides a unifying operational concept for both total and semi-synthesis of peptides and proteins.

A major debate in the origin of life is whether RNA or protein evolved first in the prebiotic age. RNA is currently favored because it possesses both enzymatic and memory functions while protein possesses only the former. However, enzymatic functions uncovered in RNAs are limited and have not been able to match the diversity and versatility displayed by proteins. Thus, it still may be useful to consider the hypothesis that proteins are evolved independently of RNA in the origin of life. How do proteins evolve without the genetic machinery? A plausible hypothesis is that they come from combinations of peptide building blocks which are derived from simple organic molecules. A part of this hypothesis requires the assembly of small peptides in aqueses solution. The orthogonal ligation, which has demonstrated its ability of free peptides to form complex proteins in water, may provide a clue to solve this puzzle.

References

1. Tam, J.P., Liu, C.-F., Lu, Y., Shao, J., Zhang, L., Rao, C. and Shin, Y.S. (1996). In Peptides: Chemistry, Structure and Biology-Proc. 14th Am. Peptide Symp. (P.T.P. Kaumaya and R.S. Hodges, Eds.) Mayflower Worldwide, Ltd., Kingswinford, UK, pp. 15.
2. Dawson, P. E., Muir, T. W., Clark-Lewis, I., and Kent, S. B. H., Science 266(1994) 776.
3. Tam, J. P., Lu, Y.-A., Liu, C. F., and Shao, J., J. Proc. Natl. Acad. Sci. USA 92(1995) 12485.
4. Baranay, G. and Merrifield, R. B., J. Am. Chem. Soc. 99(1977) 7363.
5. Kanie, O., Ito, Y. and Ogawa, T. J. Am. Chem. Soc. 116(1994) 12073.
6. Zeng, F. and Zimmerman, S. C., J. Am. Chem. Soc. 118(1996) 5326.
7. Kemp, D. S., Biopolymers 20(1981) 1793.
8. Fotouhi, N., Galakatos, N. G. and Kemp, D. S., J. Org. Chem. 54(1989) 2803.
9. Kemp, D. S. and Carey, R. I., J. Org. Chem. 58(1993) 2216.
10. Liu, C. F. and Tam, J. P., J. Am. Chem. Soc. 116(1994) 4149.
11. Liu, C. F. and Tam, J. P., Proc. Natl. Acad. Sci. USA. 91(1994) 6584.
12. Liu, C. F., Rao, C., and Tam, J. P., J. Am. Chem. Soc. 118(1996) 307.
13. Tam, J. P. and Miao, Z., J. Am. Chem. Soc. (1999) in press.
14. Tam, J. P. and Yu, Q., Biopolymer 46(1998) 319.
15. Canne, L. E., Bark, S. J. and Kent, S. B. H., J. Am. Chem. Soc. 118(1996) 5891
16. Muir, T. W., Sondhi, D. and Cole, P. A., Proc. Natl. Acad. Sci. USA 95(1998) 6705.
17. Severinov, K. and Muir, T. W., J. Biol. Chem. 273(1998) 16205.
18. Evans, T. C., Benner, J. Jr., and Xu, M.-Q., J. Biol. Chem. 274(1999) 3923.
19. Botti, P. B., Pallin, D., and Tam, J. P., J. Am. Chem. Soc. 118(1996) 10018.
20. Tam, J. P., Lu, Y.-A., and Yu, Q., J. Am. Chem. Soc. 121(1999) 4316.

Solid phase synthesis of peptide aldehydes

Jean-Alain Fehrentz[a], Marielle Paris[a], Annie Heitz[b], Marc Rolland[a] and Jean Martinez[a*]

[a]LAPP, UMR 5810 associ au CNRS-Universit Montpellier I & II, Faculté de Pharmacie, 15 av. C. Flahault, 34060 Montpellier Cedex 2; [b]UMR 9955 CNRS, Faculté de Pharmacie, 34060 Montpellier Cedex 2, France

Introduction

For the last few years, we have been involved in the synthesis of peptide aldehydes on solid support. Peptide aldehydes are excellent starting material for many chemistries (formation of reduced bond, Wittig reactions, ligation...) and have been found to be potential inhibitors of several classes of enzymes such as serine proteases [1, 2], prohormone convertases [3], cysteyl proteases [4, 5] and aspartyl proteases [6, 7]. These inhibitory properties result from the tetrahedral hydrated C-terminus aldehyde function, which mimics the transition state of the substrate during hydrolysis.

Various methods for the synthesis of peptide aldehydes in solution have been described, for example, the peptide alcohol can be oxidized into the corresponding aldehyde [8, 9]. Another widely used strategy is the diisobutylaluminium hydride reduction of the corresponding methyl esters [10]. The synthesis of argininal analogues, due to their significant activities for anticoagulant and antithrombotic properties, has been often reported using ß-lactam [2, 11, 12] and semicarbazone derivatives [1]· The synthesis of an interleukin-1 ß converting enzyme inhibitor having an aspartyl residue at the C-terminal was described by Chapman [4]. The aspartyl aldehyde moiety was protected as its corresponding O-benzylacylal which could be coupled and then hydrogenolyzed to afford the desired compound. Recent reports have mentioned the use of thiazolidines [13] or oxazolidine [14] as aldehyde precursors. However there are very few publications concerning the solid phase synthesis of peptide aldehydes [15, 16].

Results and Discussion

Our first approach based on the Weinreb amide linker [17] released the peptide aldehyde from the resin by $LiAlH_4$ reduction (fig. 1). It was efficient using Boc and Fmoc chemistry for the synthsis of short sequences without aspartyl or glutamyl residues.

The second synthesis we have described involved the reduction of phenyl esters with $AlLi(OtBu)_3H$ as reported by Zlatoidsky [18] in solution (fig. 2). The comparison of these methods and the problems of purification and racemization during purification of these products were described [19]. Until now, we were unable to synthesize the desired compounds from phenyl esters reduction without some overreduction side-reaction leading to the corresponding alcohol. This strategy can be used for the synthesis of peptide aldehydes assigned to synthesize pseudopeptides (reduced bonds, e.g.) as the peptide alcohol will not interfere in the reaction.

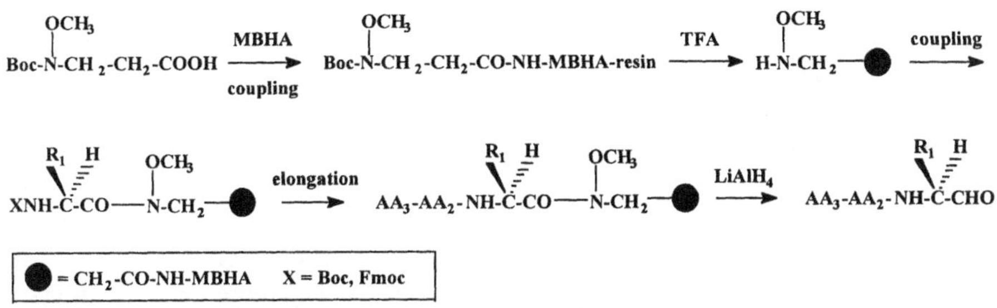

Fig. 1. Synthesis of N-protected peptide aldehydes via the Weinreb amide linker.

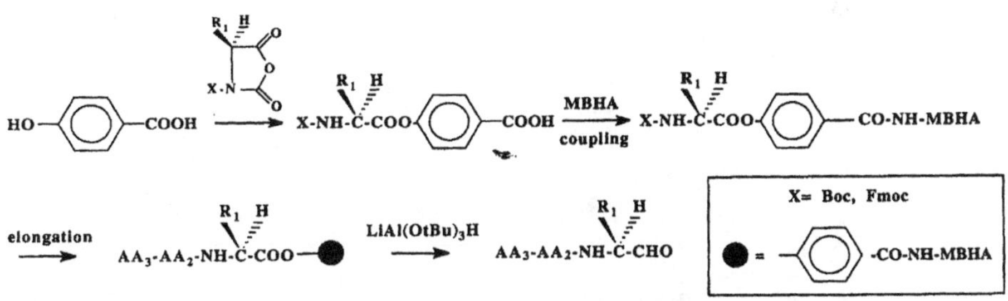

Fig. 2. Synthesis of peptide aldehydes via the phenyl ester linker.

The configuration of the C-terminal residue and its possible epimerization was studied in both strategies. As described earlier [19, 20], the aldehydic signal in ^{1}H NMR studies is a very good indicator of the possible epimerization of this residue in aldehydic peptides containing three or more residues. No epimerization occurred during these reduction reactions. We should note that no method of purification of such peptide aldehydes is known to occur without racemization. However we observed epimerization of the carbon in position □ to the aldehydic function by ^{1}H NMR spectrometry on the spectra of our model peptide aldehyde Boc-Phe-Val-Ala-H purified either by flash chromatography on silica gel as described [21] (0.1 % pyridine as eluent) or by reversed HPLC while the crude peptide aldehyde presented no epimerization. The spectra of these purified compounds revealed two aldehydic proton peaks (in $CDCl_3$) indicating that some epimerization occurred during purification. Signals corresponding to the two diastereoisomers LLL and LLD could be observed in $CDCl_3$ but not in DMSO d_6.

In order to avoid hydride reduction, which is not always compatible with side-chain groups, we proposed the use of α, ß unsaturated γ-aminoacid as a linker to the solid support to generate peptide aldehydes by ozonolysis [22]. This last methodology is very clean and proceeds without detectable racemization. The synthesis of the linker [22] was performed by a Wittig reaction between the carboethoxymethylene triphenylphosphorane and the N-protected α-amino aldehyde [23], followed by saponification to yield the corresponding ethylenic compound anchored to the solid support. After removal of the N-protecting group, elongation by classical methods of solid phase peptide synthesis (Boc or Fmoc strategies) was possible. After elongation, the peptidyl resin was subjected to an ozone stream in methylene chloride at – 80 °C and the obtained ozonide was treated with thiourea in MeOH. The peptide aldehydes were obtained with high purity and no observable racemization at the carbon in α position of the aldehyde function (fig. 3). This strategy presents several advantages: (i) aspartyl and glutamyl residues can be included in the peptide sequences without a risk of side-chain modification during the ozonolysis; (ii) ester linkage between the peptide and the resin can be used without problems (that was impossible with hydride reduction); (iii) yields are fairly good.

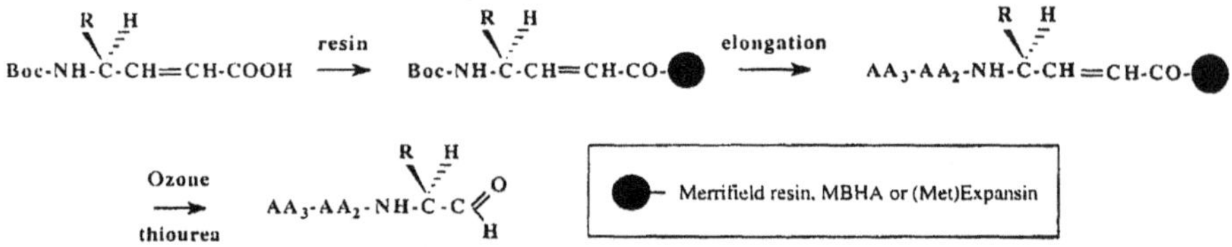

Fig. 3. Solid phase synthesis of peptide aldehydes via an α, β unsaturated γ-aminoacid linker.

The use of this linker incorporating the C-terminal residue of the peptide aldehyde implied the synthesis of a linker on the C-terminus for each different amino acid. We recently decided to explore another approach for the synthesis of peptide aldehydes by ozonolysis. This new strategy (fig. 4) consisted of the anchoring of a Wittig or Wittig-Horner reagent on the solid support, followed by the reaction with the N-protected α-amino aldehyde directly on the support. This strategy allowed the synthesis of a large amount of a general functionalized resin which is ready for the preparation of α, β unsaturated γ-aminoacyl linked to the resin from any α-amino aldehyde.

Two different approaches were tested: anchoring of diethylphosphonoacetic acid on MBHA resin with BOP as coupling reagent or anchoring of chloroacetic acid on MBHA resin with isobutylchloroformate (IBCF) as activating agent followed by reaction of the modified resin with triphenylphosphine to form the phosphonium salt. With triphenylphosphonium salts, the phosphorane is formed with butyl lithium or potassium *tert*-butylate [24]. With diethylphosphonoacetamide-resins, the carbanion is generated with various bases as described in the literature [25]. After reaction with the N-protected α-amino aldehyde, elongation of the peptide could be performed. The derivatized peptidyl resin was subjected to an ozone stream and the peptide aldehyde recovered as described previously [22]. This study was performed on our model peptide Boc-Phe-Val-Ala-H and various conditions were tested. All HPLC chromatograms of the crudes showed a high degree of purity. Surprisingly, ^{1}H NMR analysis of the aldehydic signals in the described conditions indicated epimerization of the α-carbon of the C-terminal residue (table 1). In the case of triphenylphosphonium salt, the stoichiometry of butyl lithium was decreased without any change in the observed epimerization of the resulting aldehyde and the use of potassium *tert*-butylate did not improve the reaction. Epimerization was reduced when the resin was washed by THF and DCM before the addition of the N-protected α-amino aldehyde, but the yield decreased. In this case, epimerization could be reduced to 7%.

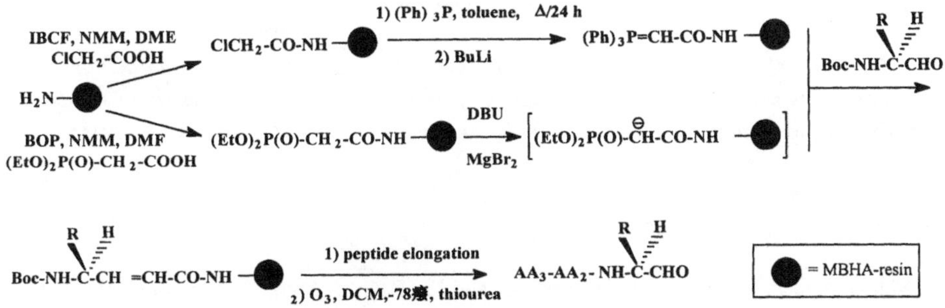

Fig. 4. Preparation of Wittig reagents linked to the solid support, peptide elongation and release of the aldehydic peptide by ozonolysis.

When diethylphosphonoacetamide was used, the situation was different since a carbanion is generated. No washing of the resin could be performed and the N-protected α-amino aldehyde was added in the presence of the base. The best conditions were found with the use of 1, 8-diazabicyclo[5.4.0]undec-7-ene (DBU) in the presence of magnesium bromide, which gave almost quantitative yields and epimerization was reduced to 9%.

These last described procedures to synthesize peptide aldehydes on solid support represent an improvement since no linker preparation was needed for each amino acid derivative. This strategy for the synthesis of peptide aldehydes on solid support allowed the preparation of large amounts of supported Witting reagent, which could then react with various carbonyl components.

In conclusion, we proposed four different approaches for the preparation of peptide aldehydes on solid support. Each strategy is an improvement in the preparation of peptide aldehydes because of the simplicity of the linker synthesis and the efficiency and rapidity of solid phase methodology. However, the strategy to be used is strongly dependent on the nature of the peptide aldehyde to be synthesized.

Table 1. Synthesis of our model peptide Boc-Phe-Val-Ala-H in various experimental conditions.

Substrate	solvent/t/time	base/equiv.	Yield (%)	HPLC purity	epimerization LLL/LLD[a]
$Cl^-,P^+(Ph)_3$-CH_2	THF/O then 65/20 h[b]	BuLi/3	90	94	73/26
"	"	*T*buOK/3	100	90	62/38
"	"	BuLi/3[c]	60	94	93/7
$(EtO)_2P(O)$-CH_2	DMF/RT/20 h	NaH/3	93	92	56/44
"	"	Et_3N, $MgBr_2$/3	57	90	89/11
"	"	DBU, $MgBr_2$/3	96	82	91/9
"	THF/RT/20 h	NaH/3	100	90	60/40
"	"	Et_3N/3	30	90	87/13

a: measured by ^1H NMR with a 360 MHz apparatus in $CDCl_3$. b: BuLi was added at 0 °C then the reaction was warmed at 65 °C. c: after the formation of the phosphorane on solid support, the resin was washed with DCM and THF before the addition of the N-protected α-amino aldehyde.

References

1. McConnell, R.M., York, J.L., Frizzel, D. and Ezell, C., J. Med. Chem., 36 (1993) 1084.
2. Bajusz, S., Szell, E., Badgy, D., Barabas, E., Horvath, G., Dioszegi, M., Fittler, Z., Szabo, G., Juhasz, A., Tomori, E. and Szilagyi, G., J. Med. Chem., 33 (1990) 1729.
3. Basak, A., Jean, F., Seidah, N.G. and Lazure, C., Int. J. Peptide Protein Res., 44(1994) 253.
4. Chapman, K.T., Bioorg. Med. Chem. Lett., 2 (1992) 613.
5. Graybill, T.L., Dolle, R.E., Helaszek, C.T., Miller, R.E. and Ator, M.A., Int. J.Peptide Protein Res., 44 (1994) 173.
6. Sarrubi, E., Seneci, P.F. and Angelastro, M.R., FEBS Letters, 319 (1992) 253.
7. Fehrentz, J.A., Heitz, A., Castro, B., Cazaubon, C. and Nisato, D., FEBS Letters, 167 (1984) 273.
8. Kawamura, K., Kondo, S., Maeda, K. and Umezawa, H., Chem. Pharm. Bull., 17 (1969) 1902.
9. Woo, J.T., Sigeizumi, S., Yamaguchi, K., Sugimoto, K., Kobori, T., Tsuji, T. and Kondo, K., BioMed. Chem. Lett., 5 (1995) 1501.
10. McConnel, R.M., Barnes, G.E., Hoyng, C.F. and Gunn, J.M., J. Med. Chem., 33 (1990) 86.
11. Balasubramanian, N., St Laurent, D.R., Federici, M.E., Meanwell, N.A., Wright, J.J., Schumacher, W.A. and Seiler, S.M., J. Med. Chem., 36 (1993) 300.
12. Schuman, R.T., Rothenberger, R.B., Campbell, C.S., Smith, G.F., Gifford-Moore, D.S. and Gesellchen, P.D., J. Med. Chem., 36 (1993) 314.
13. Galeotti, N., Plagnes, E. and Jouin, P., Tetrahedron Letters, 38 (1997) 2459.
14. Ede, N.J.and Bray, A.M., Tetrahedron Letters, 38 (1997) 7119.
15. Murphy, A.M., Dagnino, R., P.L. Vallar, Trippe, A.J., Sherman, S.L., Lumpkin, R.H.,

Tamura, S.Y. and Webb, T.R., J. Am. Chem. Soc., 114 (1992) 3156.
16. Dagnino, R. and Webb, T.R., Tetrahedron Letters, 35 (1994) 2125.
17. Fehrentz, J.A., Paris, M., Heitz, A., Velek, J., Liu, C.F., Winternitz, F. and Martinez J., Tetrahedron Letters, 36 (1995) 7871.
18. Zlatoidsky, P., Helvetica Chimica Acta, 77 (1994) 150.
19. Fehrentz, J.A., Paris, M., Heitz, A., Velek, J., Winternitz, F. and Martinez, J., J. Org. Chem., 62 (1997) 6792.
20. Fehrentz, J.A., Heitz, A. and Castro, B., Int. J. Peptide Protein Res., 26(1985) 236.
21. Ho, P.T. and Ngu, K., J. Org. Chem., 58 (1993) 2313.
22. Pothion, C., Paris, M., Heitz, A., Rocheblave, L., Rouch, F., Fehrentz, J.A. and Martinez, J., Tetrahedron Letters, 38 (1997) 7749.
23. Fehrentz, J.A. and Castro, B., Synthesis, (1983) 676.
24. Hird, N.W., Kazuyuki, I. and Nagai, K., Tetrahedron Letters, 38 (1997) 7111.
25. Blanchette, M.A., Choy, W., Davis, J.T., Essenfeld, A.P., Masamune, S., Roush, W.R. and Sakai, T., Tetrahedron Letters, 25 (1984) 3183.

Investigation of enzyme activity and inhibition in the interior of novel solid supports

Morten Meldal[a], Phaedria M.St. Hilaire[a], Marianne Willert[a], Jörg Rademann[a], Morten Grötli[a], Jens Buchardt[a], Charlotte H. Gotfredsen[a], Maria A. Juliano[b], Luiz Juliano[b] and Klaus Bock[a]

[a]*Carlsberg Laboratory, Gamle Carlsberg Vej 10, DK-2500 Valby, Denmark;* [b]*Escoloa Paulista De Medicina, Institute of Biophysics, Rua Tres de Maio 100, Sao Paulo, Brasil*

Introduction

Combinatorial chemistry has revolutionized the pharmaceutical industry, which no longer depends on the availability of natural products or existing stocks of synthetic compounds for screening in bioassays. The generation of thousands or even millions of new compounds in a single combinatorial synthetic operation, on the other hand, has created a new problem. The bioassays, which have been costly to implement, were not developed to simultaneously screen many compounds and are performed primarily in solution, requiring extensive handling for each compound assayed. The performance of assays using mixtures of compounds in solution combined with structure deconvolution [1] is not without complications. One great advantage of the solid phase split and combine method is that the compounds remain compartmentalized on separate beads [2]. If sensitive and rapid techniques can be developed for the analysis of picomolar amounts of resin bound compounds, solid phase assays could be designed for investigation of bio-molecular activities. The use of mass spectrometry, IR and solid phase magic angle spinning (MAS) NMR spectroscopy are emerging as the methods of choice in this regard.

Whereas binding assays on solid phase are readily performed with conventional polystyrene or Tentagel resins [2], the implementation of enzyme reactions on solid support is totally dependent on ready access of the enzymes to the substrates [3]. Resins containing a polystyrene core do not allow enzymes into the interior of the gel [4]. A range of PEG-based resins has therefore been developed in our laboratory. These highly polar and water-swelling resins employ long PEG chains as cross-linkers for various polymers formed by radical [5, 6], anion[6], or cation catalyzed polymerization reactions. The copolymers are composed of 80-95% of PEG, largely determining the properties of the resins. Beaded resins can be obtained by radical inverse suspension polymerizations [7] or by anion catalyzed polymerization in silicon oil. The use of polar resins for combinatorial chemistry immediately limits the types chemical reactions that can be performed because of the introduction of susceptible bonds in the polymer. However, PEG itself is quite inert and when the structure of the three-dimensional polymer network is composed of bonds with properties similar to those of PEG a stable, inert resin results. Selective reactions with masked carbanion nucleophiles have been performed in high yields on polyoxyethylene-polyoxypropylene (POEPOP) resins, and protein ligations have been reported on PEGA resins [8]. The present paper describes such gel resins and their use for enzyme reactions in the interior of the polymer matrix.

Results and Discussions

The radical polymerization reactions are performed by end group modification of the PEG with either acrylic amides by acylation of bis-2-amino-$PEG_{800, -1900, -4000 \text{ or } -6000}$ or with vinylbenzyl- or vinylphenylpropyl-ethers by alkylation of HO-PEG_{1500}-OH. The macromonomer is formed by first sodiating the PEG followed by addition of the alkyl chloride. Often some elimination occurs, and addition of more alkyl halide and base is required. The radical polymerization can in all cases be performed in inverse suspension in water/CCl_4-heptane to give beaded resins. The polymer formed with the vinylphenylpropyl-PEG_{1500} (fig. 1) is most suited for organic reactions since it contains no susceptible bonds such as amides or benzyl ethers. The latter are very unstable in the presence of strong Lewis acids, which dissolve resins containing benzylic ether bonds.

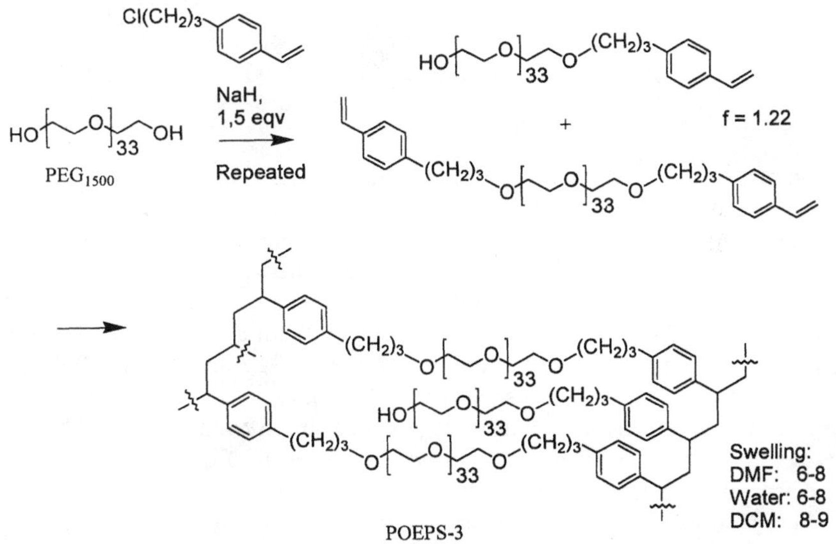

Fig. 1. Synthesis of the inert polar resin POEPS-3, which is stable to strong Lewis acids. Enzymes < ~50.000 kD diffuse freely inside the polymer network.

The reaction of epichlorohydrin with sodiated PEG yields oxirane modified PEG-molecules, which can be polymerized in the presence of potassium *t*-butoxide to afford a polymer network of ether linkages and functional hydroxyl groups. The beading in silicon oil of $POEPOP_{400}$ is achieved by droplet formation from a nozzle in a jet of nitrogen over a silicon bath of 140 °C. The droplets are allowed to polymerize as they slowly sink through the oil. Due to anion interchange during polymerization, POEPOP contains both primary and secondary ether bonds and hydroxyl groups. The presence of secondary ethers decreases the stability of the resin to a small extent and it is not stable to conditions such as HBr/AcOH or refluxing thionyl chloride. The formation of only primary ether bonds yields a significantly more stable resin, as will be described elsewhere.

The properties of the molecules linked to the solid support are very dependent on the length of the PEG macromonomer used for the polymerization. This is clearly demonstrated by ^{13}C NMR relaxation experiments with POEPOP-resins made with PEG_{400} and with PEG_{1500} (table 1). The relaxation rate of the carbon atoms in the PEG_{400} based resin is significantly higher than for the PEG_{1500} resin, indicating the reduced flexibility of the $POEPOP_{400}$ and the importance of cross-linker chain length [9].

Table 1. Relaxation rates for $PEG(CH_2)$ in PEG-based resins.

Resin	$CHCl_3$	Water
Tentagel S	1.08	2,72
POEPS	0.91	2.20
PL-PEGA	1.00	1.96
$POEPOP_{1500}$	0.95	1.76
$POEPOP_{400}$	~1.7	-

A solid phase substrate assay based on the incorporation of 2-aminobenzoic acid and nitrotyrosine into a randomized "split and combine" peptide library has been introduced by our laboratory. It was further developed into a single bead solid phase inhibitor library assay in which an inhibitor library is constructed on some of the functional groups using a base- or photo-labile linker and a substrate is attached to the rest of the functional groups. These "one-bead two compound" solid phase library assays have been used successfully to identify inhibitors (RIWRYWAV and AMMc(Bu^t)MIVF) for cruzain and subtilisin Carlsberg, respectively [10, 11].

These assays have been used with a large variety of enzymes showing the versatility of the method (table 2). The amphiphilic properties of PEG enables it to protect sensitive enzymes from denaturation and degradation. In fact, many commercial enzyme preparations are protected simply by the addition of PEG. Similarly, the enzyme activity inside the resin can be prolonged and, e.g., glycosyltransferases which often loose their activity with time are quite stable and maintain full activity over several days for the duration of the reaction.

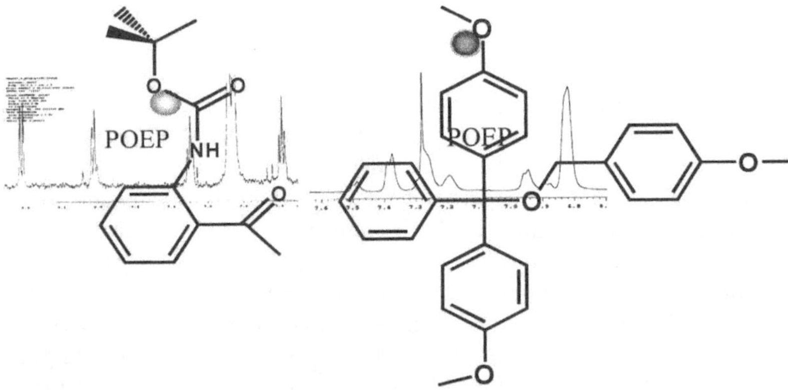

Fig. 2. MAS ^1H NMR of POEPOP resins with small molecules attached. Both the aromatic region (and the POEPOP region at 3-5.5 ppm) show little or no information of fine structure on the $POEPOP_{400}$ resin while the $POEPOP_{1500}$ show the fully resolved peaks.

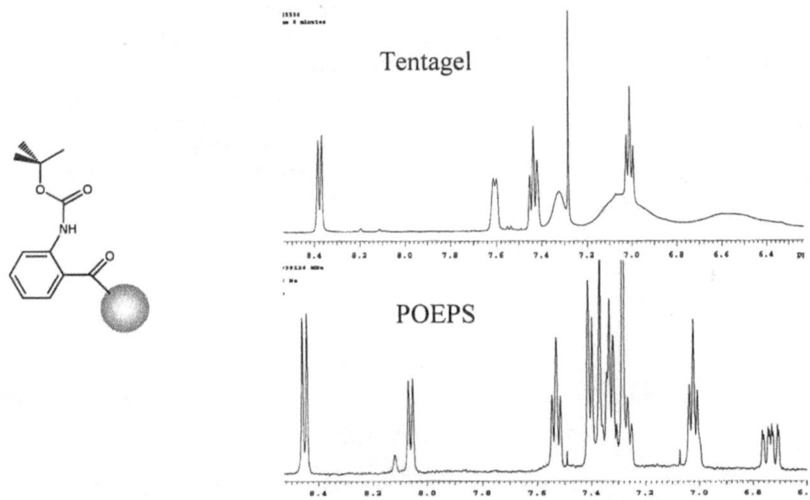

Fig. 3. The comparison of Tentagel with POEPS by MAS ^1H NMR show that the Abz-group attached is best resolved in POEPS. A remarkable difference is obse-rved for the polystyrene backbone, appearing as broad humps in Tentagel, whe-reas it is fully resolved in POEPS. Additionally the high resolution allows the id-entification of residual vinylic protons from the polymerization reaction in both resins.

Many proteases have been used successfully in PEG based resins either for reactions or in library assays (table 2). Papain, an archetypal cysteine protease, (Mw 23.4 kD; Ip

= 8.21; Net charge = +8; Specificity: very broad) and trypsin illustrate the virtues and pitfalls of the solid phase enzyme assay. A library with 7 randomized residues between resonance energy transfer donor and acceptor was constructed and tested with papain to give the substrate specificity presented in Table 3. In the first library an overrepresentation of D and E residues outside the region interacting with the active site of the enzyme was observed. This trend can only be explained by the preferential affinity of the positively charged papain to negatively charged resin beads.

The occurrence of acid residues in the binding region was, on the other hand very low. The high pseudo concentration of papain in negatively charged beads resembles a situation in which enzymes are concentrated on their substrate due to a second binding site; the presence of negative charge on lipid bilayers or a protein substrate could lead to a similar increase in local concentration. It was immediately observed that papain has a high affinity for $Y(NO_2)$ in the P2 subsite. This problem can be circumvented by insertion of a P immediately in front of the nitrotyrosine. In a second library with P inserted and D/E excluded, the full length of the active site of papain was covered.

Table 2. Enzyme reactions which have been successful on PEG based resins.

Proteases	Other enzymes
Subtilisin/PEGA, POEPS-3 or POEPOP	Fucosyl transferase/$PEGA_{4000}$
Trypsin/PEGA	β-14-Galactosyltransferase/PEGA
Chymotrypsin/PEGA	Human and Yeast PDI / $PEGA_{4000}$
Pepsin/PEGA	Unsuccessful enzyme reactions
Papain/PEGA	Sialyl transferase/PEGA
MMP9/$PEGA_{4000\text{ or }6000}$	MMP9/POEPOP, POEPS-3, $PEGA_{1900}$
MMP12/PEGA or POEPS-3	Subtilisin/$POEPOP_{400}$
Cruzipain/$PEGA_{1900}$	Cruzain, $PEGA_{300-1900}$
Leishmania-CP/PEGA	

Table 3. The substrate specificity of papain as determined by the solid phase library method.

P4	**P3**	**P72**	**P1**	**P1'**	**P2'**	**P3'**	**P4'**	**P5/6**
$Y(NO_2)$	P V	V L	A G	T A	S K	T L	N S	D/E
P F [b]	I	F M	Q K	S [a]	A [b]	K P [b]	Q C	
W Y					(D/E)		D/E	

[a] Not completely specific, [b] Preferred but not specific

Treatment of a library with different concentrations of trypsin (fig. 4), subtilisin Carlsberg, chymotrypsin and pepsin showed that most enzymes in addition to papain work similarly in the library. Long reaction times are preferred since the effect of diffusion limited kinetics is then eliminated. The reduced concentration of enzyme will increase the selectivity of the assay result.

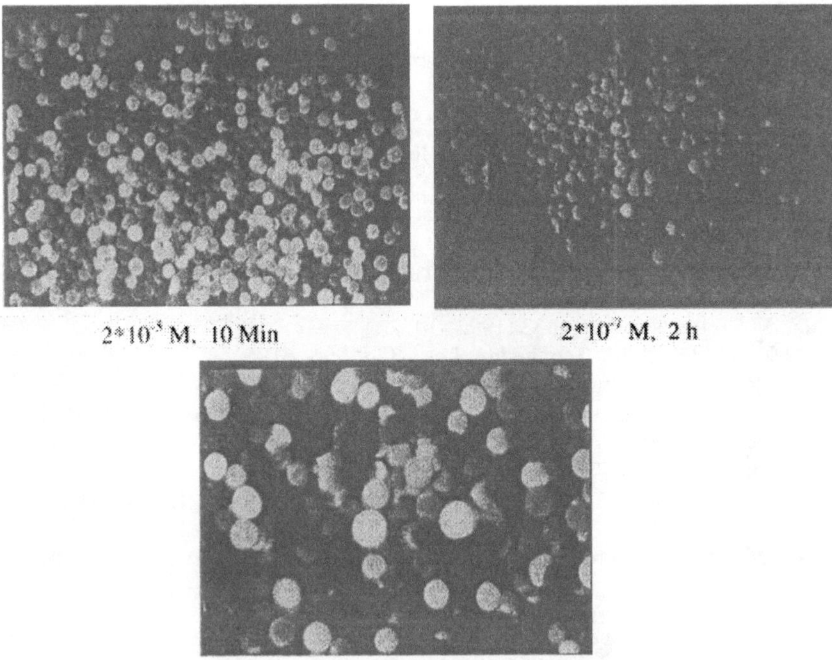

$2*10^{-5}$ M, 10 Min

$2*10^{-7}$ M, 2 h

$2*10^{-7}$ M, 24 h

Fig. 4. Treatment of XY(NO$_2$)XXXXXXXK(Abz)X-PL-PEGA$_{1900}$ with 10^{-5} M and 10^{-7} M trypsin show that selectivity of the cleavage is concentration dependent. For broad specificity enzymes like trypsin the best conditions for substrate selectivity is 24 h at a concentration of ~2×10^{-8} M.

Acknowledgements

The present work has been supported by the EU-INCO-DC program (CT970225) and by The Danish National Research Foundation.

References

1. Houghten, R.A., Pinilla, C., Blondelle, S.E., Appel, J.R., Dooley, C.T. and Cuervo, J.H. , Nature, 354 (1991) 84.
2. Lam, K.S., Salmon, S.E., Hersh, E.M., Hruby, V.J., Kazmierski, W.M. and Knapp, R.J., Nature, 354 (1991) 82.
3. Meldal, M., In Shmuel, C. (Ed.) Combinatorial Peptide Libraries, Humana Press Totowa, New Jersey, 1998, p. 51.
4. Meldal, M., In Fields, G. (Ed.) Solid-Phase Peptide Synthesis, Academic Press, 1997, p. 83.
5. Meldal, M., Tetrahedron Lett., 33 (1992) 3077.
6. Renil, M. and Meldal, M., Tetrahedron Lett., 37 (1996) 6185.
7. Arshady, R., J. Chromatogr., 586 (1991) 181.
8. Camarero, J.A., Cotton, G.J., Adeva, A. and Muir, T.W., J. Peptide Res., 51 (1998) 303.
9. (Coffey, A.F., Warner, F.P., Meldal, M. and Sørensen, M.D., Proc. of the Japanese Petide Symposium (1998), unpublished)
10. Meldal, M., Svendsen, I., Juliano, L., Juliano, M.A. , Del Nery, E. and Scharfstein, J., J. Pept. Sci., 4 (1998) 83.
11. (Meldal, M. and Svendsen, I., J. Chem. Soc. ,Perkin Trans., 1 (1995), unpublished)

Immobilization of α-Chymotrypsin on zeolite for peptide synthesis in organic solvent

Yun-Hua Ye, Guo-Wen Xing, Gui-Ling Tian and Chong-Xi Li
Department of Chemistry, Peking University, Beijing, 100871, China

Introduction

Immobilization of an enzyme usually allows the enzyme molecule to maintain its catalytic activity while other processes that are detrimental to the enzyme, such as autolysis, are inhibited. Furthermore, the immobilized enzyme is much more stable than the free enzyme during the reaction, and can be used repeatedly for a long time. In the past, various support materials were widely investigated. As an immobilization matrix, the molecular sieve has attracted great interest because of its uniform pore structure and novel properties such as high surface area, hydrophilic behavior and electrostatic interactions. In recent years, a few papers have reported the preparation of zeolite immobilized enzymes. Moreover, with the development of preparation technology, the pore diameter of zeolite can be readily changed from micropore to mesopore. There have been attempts to entrap enzymes such as papain, trypsin and cutinase on zeolite and its catalytic activity was measured [1, 2]. However, there have been no reports on using zeolite immobilized enzyme as catalyst for peptide synthesis. In a previous study we synthesized a series of peptide derivatives including N-protected Leu-enkephalin by enzyme in organic solvent [3-5]. In this work, a new type of immobilized α-chymotrypsin with different kinds of zeolites as matrixes was used for the first time, and a tripeptide derivative, Z-Tyr-Gly-Gly-Oet, was successfully synthesized from Z-Tyr-OEt and Gly-Gly-OEt in dichloromethane. The effect of zeolite properties on enzymatic synthesis was studied. Some reaction conditions in the synthesis were compared.

Results and discussion

Preparation of immobilized α-chymotrypsin on zeolite

10 mg α-chymotrypsin (type II, from Sigma) was dissolved in 3.3 ml phosphate buffer (pH 7.73, 50 mM), and then 100 mg zeolite was added at room temperature with stirring. The filtrate from the suspension solution was taken every 10 minutes, and the activity of α-chymotrypsin was spectrophotometrically measured with acetyl tyrosine ethyl ester (ATEE) as substrate [6]. After about an hour the activity of α-chymotrypsin in the filtrate decreased to about zero, suggesting that the α-chymotrypsin molecules were almost completely adsorbed on the zeolite. The suspension solution was then lyophilized to obtain the immobilized α-chymotrypsin.

Synthesis of ZTyrGlyGlyOEt by immobilized α-chymotrypsin

ZTyrOEt (0.5 mmol) and HCl·GlyGlyOEt (0.5 mmol) were suspended in 5 ml dichloromethane, and triethylamine (1.0 mmol) was added. After 50mg immobilized α-chymotrypsin and water (12.5 μl, H_2O/CH_2Cl_2=0.25% (V/V)) were added, the mixture was stirred at room temperature for 1-3 days. The reaction was detected by TLC. At the end of the reaction, the precipitate from the reaction solution and the immobilized enzyme were filtered and washed thoroughly by acetone. Then the immobilized α-chymotrypsin was dried in vacuum for storage and reuse, and the acetone solution was evaporated to obtain ZTyrGlyGlyOEt, m.p. 164-165 °C, $[\alpha]_D^{20}$ +2.9 (c 2 HOAc), yield 45%-69% (first use). In literature [7], m.p. 169-171 °C, $[\alpha]_D^{20}$ +8.1 (c 1 MeOH).

Reusability of immobilized α-chymotrypsin

Fig. 1 illustrates the relationship between peptide yield and the reuse times of the zeolite immobilized enzyme. The results show that α-chymotrypsin immobilized by physical adsorption over different zeolites ranging from microporous molecular sieves (HY, NH_4Y, NaY) to mesoporous ones (HDAY, HNH_4DAY, MCM-41, DAY stands for dealuminized Y zeolite) was active for peptide synthesis. HY zeolite was the best matrix since the tripeptide was obtained in 45% yield after the immobilized enzyme was used 6 times.

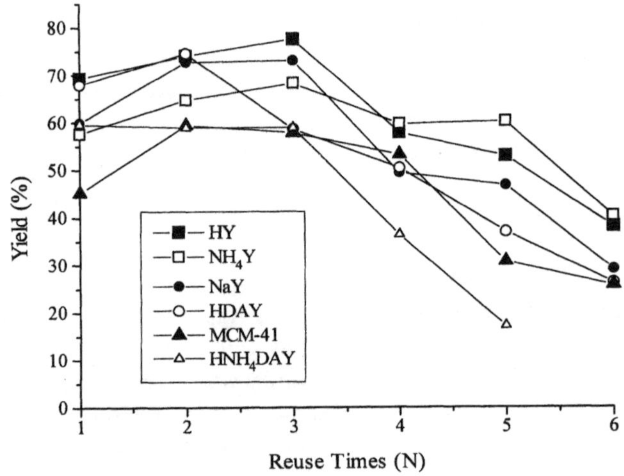

Fig. 1. The reusability of immobilized α-chymotrypsin in the synthesis of Z-Tyr-Gly-Gly-OEt. HY stands for the HY zeolite, NH_4Y, HDAY, HNH_4DAY and NaY are the same.

Although the pore sizes of dealuminized Y zeolite and MCM-41 are larger than that of Y zeolite and can offer the possibility of accommodating enzyme molecules into the cages or channels, the yield of model peptide was not higher than that with microporous HY zeolite as matrix. The maximum yield of Z-Tyr-Gly-Gly-OEt was 77.6% when the HY immobilized α-chymotrypsin was used three times for enzymatic peptide synthesis.

Effect of the structure property of zeolite

It is suggested that the H-bond acceptors on the molecule surface of the α-chymotrypsin can form H-bonds with the hydroxyl groups on the zeolite. In the case of HY zeolite, there are many -OH outside the pore opening on the external surface of zeolite. The HY zeolite uses its OH groups to form tight H-bonds with those H-bond acceptors on the surface of the enzyme. The complex structure is very stable. Therefore the enzyme can catalyze the peptide bond formation repetitively. Pore widening of the zeolite to medium size (HDAY as an example) or large size (MCM-41) does not enhance the catalytic activity of the enzyme because the process of pore widening reduces the number of OH groups on the zeolite surface. Thus, without many hydroxyls on the surface, the enzyme comes off easily. In a word, the catalytic activity of the immobilized enzyme depends greatly on the strength of the H-bond of the zeolite-enzyme complex. It has little to do with the pore size of the zeolite. Study of the nature of the H-bonding and the complex structure is currently underway.

Table 1. The effect of water content of dichlorometane on the product yield. HY immobilized α-chymotrypsin as catalyst.

Water content %(V/V)	0	0.15	0.25	0.50	0.75	1.00
Product yield %	29.7	66.0	69.3	76.4	74.2	77.6

Table 2. The reusability of HY immobilized α-chymotrypsin under different water content.

Water content %(V/V)	Product yield in different reuse times %			
	1	2	3	4
0.25	69.3	74.1	77.6	57.8
1.00	77.6	77.3	37.6	11.2

Effect of water content of dichloromethane

The water content of organic solvent greatly influenced the activity of immobilized enzyme and the product yield. With HY immobilized α-chymotrypsin as an example, different amounts of water were added to the reaction. The water content of dichloromethane was varied from 0 to 1.0% (V/V). Table 1 shows that when no water was injected into the media, the model peptide was obtained in low yield, but with increasing water the product yield rapidly improved. However when the water content increased to ≥0.50% (V/V) the yield leveled off. Molecular sieves can adsorb a small amount of water during lyophilization which serves as the essential water of enzyme in organic solvent. Thus, to a certain degree, the HY immobilized α-chymotrypsin in dry dichloromethane can show activity even without water added. Although the HY immobilized α-chymotrypsin possessed higher catalytic activity, when more water was added to the solvent its reusability decreased remarkably as indicated in Table 2. Under the condition of 0.25% water content, the HY immobilized α-chymotrypsin can be efficiently used several times with a product yield of 57.8% in the 4th reuse. In comparison, when 1.0% water content was chosen as the reaction parameter, the HY immobilized α-chymotrypsin could be used effectively only 2 times. With the third reuse, the activity of immobilized enzyme declined and the tripeptide was obtained in poor yield. These showed that the increase of water concentration can on the one hand, promote the catalytic efficiency of immobilized enzyme, and on the other hand, greatly deactivate the enzyme molecules. Therefore, to improve the reusability of immobilized enzyme and maintain high product yield, it is necessary to add the proper amount of water such as 0.25% (V/V) to the solvent.

Conclusion

We used zeolite immobilized α-chymotrypsin for peptide synthesis in dichloromethane. HY, NH_4Y, NaY, HNH_4DAY, HDAY and MCM-41 zeolites with different acidity and different pore sizes were compared with the synthesis of ZTyrGlyGlyOEt as a model reaction. The reusability of the six different zeolites was studied, and the results showed that HY zeolite was the best matrix for immobilization of α-chymotrypsin; dealusminized Y zeolites and MCM-41 were not better supports for enzymatic reaction than Y zeolites. It was suggested that the activity of immobilized α-chymotrypsin depended greatly on the strength of the H-bond of the zeolite-enzyme complex, and had little to do with the pore dimension of the zeolite. The activity of immobilized enzyme and product yield were influenced significantly by the water content of dichloromethane. A little water around the enzyme molecules is truly necessary to maintain its activity. There are also many questions raised about the reaction such as how to identify the H-bond formation between enzyme and zeolite, and how to hold the enzyme tightly to zeolite. All of the above problems need to be studied further.

Acknowledgments

This work was supported by Doctoral Program Foundation of Institution of Higher Education. The authors are grateful to Professor Xuan-Wen Li for supplying all zeolites and helpful discussion.

References

1. Goncalves, A.P.V., Lopes, J.M., Lemos, F., Ramoa Ribeiro, F., Prazeres, D.M.F., Cabral, J.M.S. and Aires-Barros, M.R., J. Molecular Catalysis B: Enzymatic,1 (1996) 53.
2. Diaz, J.F. and Balkus Jr, K.J., Joural of Molecular Catalysis B: Enzymatic, 2 (1996) 115.
3. Tian, G.L., Liu, Y., Wang, H. and Ye, Y.H., Chem. J. Chin. Univ., 17 (1996) 55.
4. Xing, G.W., Tian, G.L., Lu, Y. and Ye, Y.H., In Xu, X.J., Ye, Y.H. and Tam, J.P.(Eds.) Peptides: Biology and Chemistry (proceedings of the 1996 Chinese Peptide Symposium), Kluwer Academic Publishers, The Netherlands, 1998, p. 50-51.
5. Ye, Y.H., Tian, G.L., Xing, G.W., Dai, D.C., Chen, G. and Li, C.X., Tetrahedron, 54 (1998) 12585.
6. Zhang, L.X., Zhang, T.F. and Li, L.Y., the Experimental Technology and Method for Biological Chemistry, the People's Education Press, Beijing, 1981, p. 140-141.
7. Gill, I. and Vulfson, E.N., J. Chem. Soc. Perkin Trans., 1 (1992) 667.

Preparation of a peptide thioester using a fluoren-9-ylmethoxycarbonyl solid-phase method

Xiang-Qun Li, Toru Kawakami and Saburo Aimoto
Institute for Protein Research, Osaka University, 3-2 Yamadaoka, Suita, Osaka, 565-0871, Japan

Introduction

The thioester method, in which partially protected peptide *S*-alkyl thioesters are used as building blocks, has been shown to be useful for polypeptide syntheses [1-3]. Typically, such building blocks are prepared by a Boc solid-phase method [4, 5]. Although the fluoren-9-ylmethoxycarbonyl (Fmoc) solid-phase method [6, 7] is widely accepted for peptide synthesis, it cannot be employed for the preparation of peptide thioesters, since thioesters are easily decomposed by aminolysis during the removal of Fmoc groups by treatment with piperidine. The Fmoc solid-phase method requires neither the repetitive use of trifluoroacetic acid nor strong acid treatment in order to obtain free peptides. These features are advantageous not only to the preparation of simple peptides, but also for the preparation of conjugated peptides such as phosphopeptides and glycopeptides. As a result, a new route for the preparation of a wide variety of polypeptides by the thioester method and chemical ligation methods [8-11] is possible, provided suitable conditions for the preparation of peptide thioesters by the Fmoc solid-phase method can be developed.

Results and Discussion

We therefore carried out detailed studies of conditions for the removal of Fmoc groups while keeping the thioester intact. Using a model peptide thioester, Fmoc-Phe-Leu-Ala-Cys(Acm)-His-Gly-$SCH_2CH_2CONH_2$ (**1**), we evaluated the effect of a series of amines on the rates of the removal of Fmoc groups and the aminolysis of thioester moieties. Piperidine, cyclohexylamine, 4-aminomethyl-piperidine and morpholine completely removed the Fmoc group, but the thioester moiety was also completely decomposed by aminolysis. Hexamethyl-eneimine and heptamethyleneimine completely removed the Fmoc group while 20 to 25% of the thioester moieties remained intact. Sterically hindered amines, such as cis-2, 6-dimethylpiperidine, dicyclohexylamine and diisopropylethylamine, did not remove the thioester moiety, but were not sufficiently strong to remove the Fmoc group within 20 min. Although neither 1-methylpiperidine nor 1-methylpyrrolidine destroyed the thioester moiety, the Fmoc group was not removed by 1-methylpiperidine even after one hour treatment. However, 1-methylpyrrolidine removed the Fmoc group completely within 20 min. HOBt was added to the reaction mixture to lower the nucleophilicity of the amines, because HOBt is known to suppress the cyclization of the aspartyl residue during base treatment [12,13]. In our hands, HOBt suppressed the aminolysis of the thioester. The *S*-tertiary alkyl thioester is slightly more stable than *S*-primary alkyl thioester.

Taking these results into account, the solid-phase synthesis of the partial sequence of Vero-toxin [14] was carried out according to the procedures of Scheme 1. to establish conditions for the practical solid-phase synthesis of peptide thioesters by Fmoc strategy. The crude product obtained after treatment with Reagent K[15] was analyzed by reversed-phase HPLC and each isolated peak was subjected to MALDI-TOF mass analysis to characterize the product. The yields of the desired product, H-Tyr-Thr-Lys-Tyr-Asn-AspAsp-Asp-Thr-Phe-Thr-Val-Lys-Val-Gly- $SC(CH_3)_2CH_2$ $CONH_2$ [H-Vero-toxin (11-25)-$SC(CH_3)_2CH_2CONH_2$ (3)] were calculated based on the Gly content in the starting resin and quantitative amino acid analysis, and are summarized in Table 1.

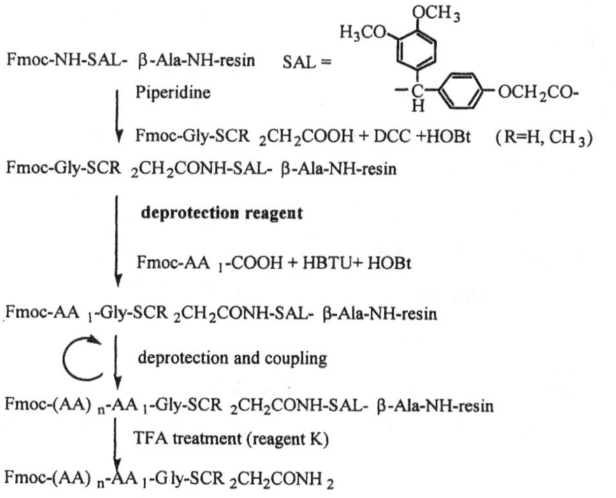

Scheme 1. Procedures for p reparation of a peptide thioester by a *n Fmoc SPPS.*

No desired product was obtained when NMPontainin solutions cg piperidine (25%, v/v), or both piperidine (25%, v/v) and HOBt (2%, w/v) were used as deblocking reagents. All the deblocking reagents, including NMP solutions containing hexamethyleneimine (25%, v/v) with or without HOBt (2%, w/v) and NMP solutions containing N-methylpyrrolidine (25%, v/v) with or without HOBt gave poor results. The yields of the desired product were low and numerous by-products were observed. As a result, a three-component deblocking reagent was prepared, which contained *N*-methylpyrrolidine, hexamethyleneimine and HOBt. This reagent greatly improved the yields and was used as a deblocking reagent. The yield was further improved by the use of a 1:1 mixture of NMP and DMSO as a solvent to dissolve the three components.

These results suggest that this mixture efficiently solvated the growing peptide chain [16], that *N*-methylpyrrolidine was a sufficiently strong base to remove Fmoc groups, that hexamethyleneimine was an efficient nucleophile for scavenging the reactive dibenzofulvene, and that HOBt efficiently suppressed the aminolysis caused by hexamethyleneimine.

This deblocking reagent, which contains 1-methylpyrrolidine (25% v/v), hexamethyleneimine (2% v/v) and HOBt (2% w/v) in a 1:1 mixture of NMP-DMSO, was applied to the synthesis of Thr-Pro-Asp-Cys(Acm)-Val-Thr-Gly-Lys- Val-Glu-Tyr-Thr-Lys-Tyr-Asn-Asp-Asp-Asp-Thr-Phe-Thr-Val-Lys-Val-Gly-$SC(CH_3)_2CH_2CONH_2$ ([Cys(Acm)[4]]-Vero-toxin (1-25)-$SC(CH_3)_2$ CH_2 $CONH_2$) (fig. 1). The peptide thioester was obtained in a yield of 22%, slightly higher than the 19% yield obtained by Boc strategy.

Table 1. Synthesis of Vero-toxin(11-25)-$SC(CH_3)_2CH_2CONH_2$ using different deblocking reagents.

deblocking reagent	yield of desired product (%)
piperidine (25% v/v) / NMP	0
piperidine (25% v/v), HOBt (2% w/v) / NMP	0
Hexamethyleneimine (25% v/v) / NMP	1.9
Hexamethyleneimine (25% v/v), HOBt (2% w/v) / NMP	3.3
1-methylpyrrolidine (25% v/v) / NMP	5.0
1-methylpyrrolidine (25% v/v), HOBt (2% w/v) / NMP	6.8
1-methylpyrrolidine (25% v/v), hexamethylene imine (2% v/v), HOBt (2% w/v) /NMP	14
1-methylpyrrolidine(25% v/v), hexamethylene imine (2% v/v), HOBt (2% w/v) / NMP-DMSO (1:1)	24

The deblocking reagent we developed herein represents a significant improvement in the synthesis of peptide thioesters in terms of both yield and quality of crude products. This represents an important step for the generalization of the thioester method and the preparation of conjugated polypeptides, such as phosphoproteins and glycoproteins.

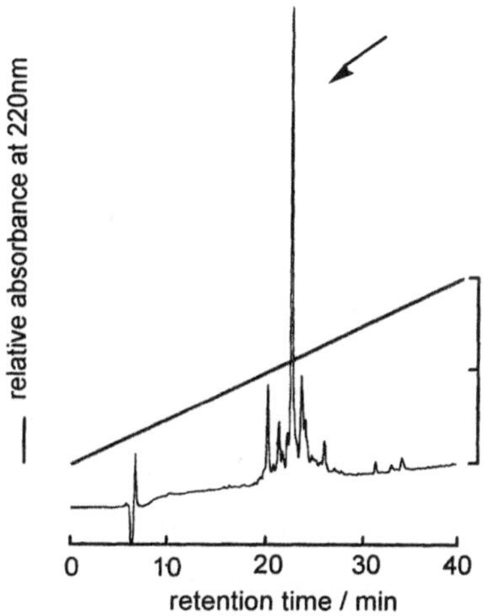

Fig. 1. An analytical HPLC elution profile of crude. ([Cys(Acm)^4]-Vero-toxin(1-25)-$SC(CH_3)_2CH_2CONH_2$. Reservoir A contained acetonitrile containing 0.1% TFA and reservoir B contained 0.1% asq. TFA. Peptides were analyzed by a linear gradient of 10 - 50% A over 40 min using Cosmosil $5C_{18}AR$ (10 × 250 mm) at a flow rate of 2.5 ml/min. An arrow indicates the desired product.

References

1. Hojo, H. and Aimoto, S., Bull. Chem. Soc. Jpn., 64 (1991) 111.
2. Hojo, H., Yoshimura, S., Go, M. and Aimoto, S., Bull. Chem. Soc. Jpn., 68 (1995) 330.
3. Kawakami, T., Kogure, S. and Aimoto, S., Bull. Chem. Soc. Jpn., 69 (1996) 3331.
4. Merrifield, R.B., J. Am. Chem. Soc., 85 (1963) 2149.
5. Kent, S.B.H., Annu. Rev. Biochem., 57 (1988) 957.
6. Atherton, E., Fox, H., Harkiss, D., Logan, C. J., Sheppard, R.C. and Williams, B.J., J. Chem. Commun., (1978), 537.
7. Chang, C.D. and Meienhofer, J., Int. J. Peptide Protein Res., 11 (1978) 246.
8. Dawson, P.E., Muir, T.W., Clark-Lewis, I. and Kent, S.B.H., Science, 266 (1994) 776.
9. Canne, L.E., Bark, S.J. and Kent, S.B.H., J. Am. Chem. Soc., 118 (1996) 5891.
10. Hackeng, T., Mounier, C.M., Bon, C., Dawson, P.E., Griffin, J.H. and Kent, S.B.H., Proc. Natl. Acad. Sci. USA, 94 (1997) 7845.
11. Tam, J.P., Lu, Y.A., Liu, C.F. and Shao, J., Proc. Natl. Acad. Sci. USA, 92 (1995) 12485.
12. Martinez, J. and Bodanszky, M., Int. J. Peptide Protein Res., 12 (1978) 277.
13. Lauer, J.L., Cynthia, C., Fields, G. and Fields, G.B., Lett. Peptide Science, 1 (1994) 197.
14. Seidah, N.G., Donohue-Rolfe, A., Lazure, C., Auclair, F., Keusch, G.T. and M. Chretien, J. Biol. Chem., 261 (1986) 13928.
15. King, D. S., Fields, C.G. and Fields, G.B., Int. J. peptide protein Res., 36 (1990) 255.
16. Fields, G.B. and Fields, C.G., J. Am. Chem. Soc., 113 (1991) 4202.

Application of DEPBT for the synthesis of N-protected peptide alcohols

Gui-Ling Tian, Bin-Yuan Sun, Wei Wang and Yun-Hua Ye
Department of Chemistry, Peking University, Beijing, 100871, China

Introduction

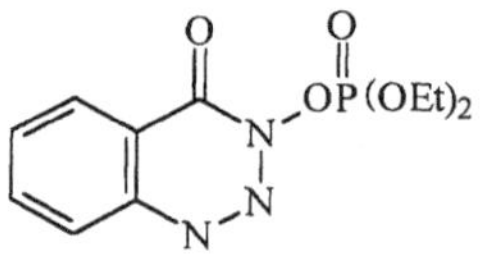

DEPBT

3-(diethoxyphosphoryloxy)-1,2,3-benzotriazin-4(3H)-one (DEPBT) is an organophosphorus compound developed by our group which can be used as an efficient coupling reagent for peptide synthesis [1]. Recent studies have demonstrated that DEPBT can be used in either solution-phase or solid-phase synthesis [2]. The advantages of DEPBT are that it has good solubility and causes less than 1% racemization. N-protected amino alcohols and N-protected peptide alcohols have attracted attention, as important synthetic intermediates. Amino alcohols can be used as chiral ligands to catalyze asymmetric catalytic reactions. In this paper DEPBT is used for the first time in the synthesis of peptide alcohols. Our experimental results demonstrated that DEPBT can be used not only in peptide synthesis, but also in the synthesis of amide between amino acid and amino alcohol. DEPBT selected the couplings of amino groups in amino alcohols and the hydroxy groups need not be protected.

Results and Discussion

Threoninol and serinol were prepared by reducing the mixed anhydrides of N-protected threonine and serine with $NaBH_4/H_2O$. This method is convenient because the hydroxy group is not protected. Fmoc-Cys(Acm)-Thr(ol), Fmoc-D-Phe-Thr(ol), Fmoc-Trp-Thr(ol), Z-Tyr(OBzl)-Thr(ol), Fmoc-Cys(Acm)-Ser(ol), Fmoc-D-Phe-Ser(ol), Z-Tyr(OBzl)-Ser(ol) and Fmoc-Lys(Boc)-Thr-Cys(Acm)-Thr(ol) were synthesized successfully by DEPBT as a coupling reagent in pH=8-9 (scheme 1). Their structures were confirmed by MS and elemental analysis. Their physical properties are listed in Table 1.

Conclusion

A convenient reduction method of threonine and serine without the protection of their hydroxy groups was developed by our group. Thr(ol) and Ser(ol) were readily prepared in yields of 65% and 68%, respectively, from Boc-ThrOH and Boc-SerOH [3].

$$\text{P-AAOH} + \text{H}_2\text{N-}\underset{\text{R}}{\underset{|}{\text{CH}}}\text{-CH}_2\text{OH} \xrightarrow[\text{Et}_3\text{N}]{\text{DEPBT}} \text{P-AA-NH-}\underset{\text{R}}{\underset{|}{\text{CH}}}\text{-CH}_2\text{OH}$$

(1) (2)

(1): Fmoc-Cys(Acm)OH, Fmoc-D-PheOH,
Fmoc-TrpOH, Z-Tyr(OBzl)OH.

(2): Thr(ol), Ser(ol).

Scheme 1. The synthesis of peptide alcohol derivatives.

Table 1. The physical properties of peptide alcohols using DEPBT as coupling reagent.

No	Acyldonor	Nucleophile	Product	Yield (%)	mp. (°C)*	$[\alpha]_D^{20}$ **
1	Fmoc-TrpOH	Thr(ol)	Fmoc-Trp-Thr(ol)	60	131-134	-23.4
2	Z-Tyr(OBzl)OH	Thr(ol)	Z-Tyr(OBzl)-Thr(ol)	76	126-129	-14.4
3	Fmoc-D-PheOH	Thr(ol)	Fmoc-D-Phe-Thr(ol)	48	174-175	+23.0
4	Fmoc-Cys(Acm)OH	Thr(ol)	Fmoc-Cys(Acm)-Thr(ol)	68	112-115	-10.9
5	Fmoc-Cys(Acm)OH	Ser(ol)	Fmoc-Cys(Acm)-Ser(ol)	47	107-109	-29.1
6	Fmoc-D-PheOH	Ser(ol)	Fmoc-D-Phe-Ser(ol)	67	129-131	+15.7
7	Z-Tyr(OBzl)OH	Ser(ol)	Z-Tyr(OBzl)-Ser(ol)	77	147-149	-11.8
8	Fmoc-Lys-(Boc)-ThrOH	H-Cys(Acm)-Thr(ol)	Fmoc-Lys(Boc)-Thr-Cys(Acm)-Thr(ol)	69	142-144	-10.5

** Melting points were taken on Yanaco macro melting point apparatus and uncorrected.*
*** C=1.0 (except Fmoc-Trp-Thr(ol), C=0.32); Solvent: DMF.*

Because Thr(ol) and Ser(ol) have two hydroxy groups and one amino group in each molecule, they have very good solubility in water. It was found that some N-Boc-protected dipeptide alchohols have very good solubility in water. To facilitate the separation of the products, the α-amino group can be protected better by Fmoc than by Boc.

Fmoc-Cys(Acm)-Thr(ol) was synthesized by the DEPBT method and the mixed anhydride method. The products obtained by these two methods has the same physical properties.

Our present work showed that DEPBT could be used not only in peptide synthesis, but also in the synthesis of peptide alcohols. When DEPBT is used as a coupling reagent for synthesis of peptide alcohols, the peptide bond is formed by the amino group of the amino alcohol, the hydroxy group of amino alcohol did not react with acyldonor, and no ester bond was produced. DEPBT has good selectivity to the amino group in amino alcohols, but DCC has no selectivity in the synthesis of peptide alcohols.

Acknowledgments

The authors thank the National Natural Science Foundation of China for financial support.

References

1. Fan, C.X., Hao, X.L. and Ye, Y.H., Synthetic Communications, 26 (1996) 1455.
2. Ye, Y.H., Fan, C.X., Zhang, D.Y., Xie, H.B., Hao, X.L. and Tian, G.L., Chemical Journal of Chinese Universities, 18 (1997) 1086.
3. (Sun, B.Y., Tian, G.L., Wang, W., Jin, N. and Ye, Y.H., Acta Scientiarum Naturalium, Universitatis Pekinensis, 1998, unpublished).

Decomposition of amino acid cupric complex by using tetrahydrothiazole-2-thione for preparation of N^{α}-Boc-N^{ε}-Fmoc-L-Lysine

Xing-Ming Gao, Yun-Hua Ye, Gui-Ling Tian and Ding Wang
Department of Chemistry, Peking University, Beijing, 100871, China

Introduction

In the synthesis of peptides containing ornithine, lysine or tyrosine, it is necessary to protect their side chain functional groups. This can be accomplished by reacting the α-amino and carboxyl groups of these amino acids with cupric ions to form stable chelate complexes. The side chain functional groups are then selectively protected, and finally, the cupric complexes are decomposed to obtain the desired protected amino acids side chains. Various compounds such as H_2S, EDTA, HCl and thioacetamide, etc. have been used as decomposing reagents of cupric complexes [1, 2]. However, H_2S is highly toxic and has an unpleasant smell; besides, the side product (CuS) sometimes is difficult to filter due to its colloidal form. The cupric complexes are not easy to decompose completely by EDTA and require long reaction times. Treatment with HCl cannot be used for N^{δ}-Z-L-Orn or N^{ε}-Z-L-Lys because they are soluble in acids. The method of thioacetamide for the preparation of O-Bzl-L-Tyr results in low yield.

In previous papers, we reported that tetrahydrothiazole-2-thione (TTT) was readily reacted with amino acids to obtain active amides for peptide synthesis [3]. In addition, TTT was also used to decompose cupric complexes for preparation of N^{ε}-Z-L-Lys and O-Z-L-Tyr [4]. The results were much better than those from H_2S and EDTA methods.

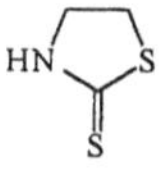

TTT

In this article, we report decomposition of lysine cupric complex using TTT for preparation of N^{a}-Boc-N^{e}-Fmoc -L-Lysine in a good yield.

Results and discussion

As shown in Scheme 1, lysine was reacted with $CuCO_3 \bullet Cu(OH)_2$ to form a cupric complex and then coupled with N-(fluorenylmethoxycarbonyloxy)succinimide (Fmoc-OSu) to provide $Cu[Lys(Fmoc)]_2$. The cupric complex was successfully decomposed by TTT. In order to compare the decomposing efficiency, H_2S, EDTA and HCl were also studied in the same reaction.

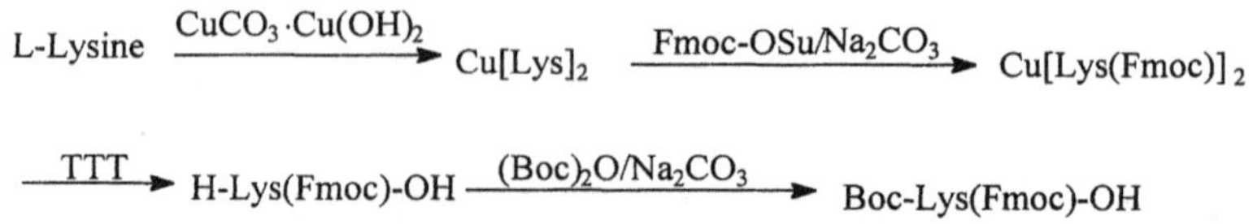

Scheme 1.

The decomposing procedure was performed as follows: TTT was added to the methanol solution containing Cu[Lys(Fmoc)]$_2$ with stirring. The yellow solid (Cu(II)-TTT complex) formed in the solution was filtered and discarded. The filtrate was concentrated to get the crude product. Recrystallization from methanol-water gave pure N^{ε}-Fmoc-L-Lys.

The results are listed in Table 1.

Table 1. Comparison of decomposition of Cu[Lys(Fmoc)]$_2$ by different reagents.

Reagents	m.p. (°C)	Solvents	Time	Yield (%)
TTT	191-194	MeOH	5 min	80
H_2S	190-193	MeOH	8 hr	86
EDTA	208-211	H_2O	48 hr	75
HCl	192-194	H_2O	30 min	62

The amino acid cupric complex was decomposed completely by TTT in a few minutes with good yield. The product was easy to purify since N^{e}-Fmoc-L-Lys was soluble in methanol while the Cu(II)-TTT complex precipitated from methanol. The results showed that the decomposing efficiency of TTT was higher than that of EDTA. Moreover, the TTT method avoided the unpleasant smell caused by H_2S. In summary, TTT is a good reagent for the preparation of protected side chain derivatives of tyrosine and lysine from their cupric complexes. Further studies are in progress.

When N^{ε}-Fmoc-L-Lys was reacted with $(Boc)_2O$ in aqueous Na_2CO_3, N^{α}-Boc-N^{ε}-Fmoc-L-Lys was obtained in 74% yield. m.p. 75-78°C, $[\alpha]_D^{20} = -1.4$ (c 1,MeOH), MS-FAB:(m/z) = 469 $(M+H)^+$, Analysis for calculated $C_{26}H_{32}N_2O_6$ (%): C, 66.65; H, 6.88; N, 5.98, Found (%):C, 66.87; H, 7.14; N, 6.08.

Acknowledgments

The authors thank the National Natural Science Foundation of China for financial support.

References

1. Taylor, U., Dyckes, D.F. and Cox, J.R., Int. J. Peptide Protein Res., 19 (1982) 158.
2. Albericio, F., Nicolas, E., Rizo, J., Ruiz-Gayo, M. and Pedroso, E., Synthesis, (1990) 119.
3. Li, C.X., Ye, Y.H., Lin, Y. and Xing, Q.Y., Tetrahedron Lett., 22 (1981) 3467.
4. Ye, Y.H., Fu, G.Q., Yu, N. and Li, C.X., Kexue Tongbao, 30 (1985) 1873.

Two-step selective formation of three disulfide bridges in the synthesis of δ-Conotoxin PVIA

Hui Jiang[a], Chong-Xu Fan[a], Zheng-Wei Miao[b], Kai-Hua Wei[a], Da-Yu Li[a], Ming-Nai Zhong[a] and Ji-Sheng Chen[a]

[a]Research Institute of Pharmaceutical Chemistry, Beijing, 102205; [b]Department of Chemistry, Peking University, Beijing, 100871, China

Introduction

Strategies to create multidisulfide regioselectively in synthetic peptides have recently been reviewed [1, 2], including the single step oxidative method and the step by step oxidative method [3]. In the first approach, disulfide bridges are formed by random oxidation, usually leading to undesired disulfides; there are 15 possible arrangements of disulfide bridges with 6 cysteines in single-chain. The second approach is to form disulfide sequentially. This is a difficult undertaking because of the need for three orthogonal pairs of thiol protecting groups and poor yield. A two-step strategy (two disulfides formed first, then the third) has been utilized to treat the three-disulfide containing peptide [4]. But the choice of which pair of cysteines to block in the first step is important. Three possible products form after the first step oxidation, only one of which produces a properly folded product. A coincident and difficult problem is the separation and identification of the isomers, also, the location of the position of the three disulfides in a given isomer can be particularly challenging.

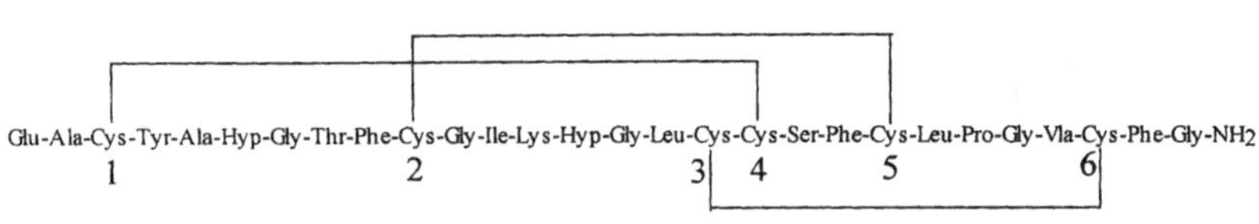

Fig.1. δ-conotoxin with Cys residues numbered consecutively from N- to C-terminus. Native toxin pairings are 1-4, 2-5, and 3-6. The Cys framework is as the same as ω- conotoxins.

In this report we designed a new two-step oxidation approach: to form one disulfide in the first step, while the remaining disulfides are formed in the second step. We chose δ–conotoxin PVIA [5], a 3- disulfide, 29-residue peptide to carry out a series of folding studies (fig. 1).

Results

Synthetic approach

The peptides were synthesized using Boc chemistry with stepwise solid-phase technology. The six cysteines were selectively protected with two different blocking-groups: four with acetamidomethyl (Acm), and two with 4-methylbenzyl (Mbzl). In the first step, liquid HF was used to cleave the peptide from the resin and simultaneously deblock all the side chain protecting groups except the Acm moiety. Thus, two cysteines were free and oxidized by DMSO [6] to form one- disulfide bridge without isomers. In the second oxidation process, I_2 was used to remove the Acm groups to form two disulfides. This approach reduces the number of possible disulfide bridging patterns from 15 to 3.

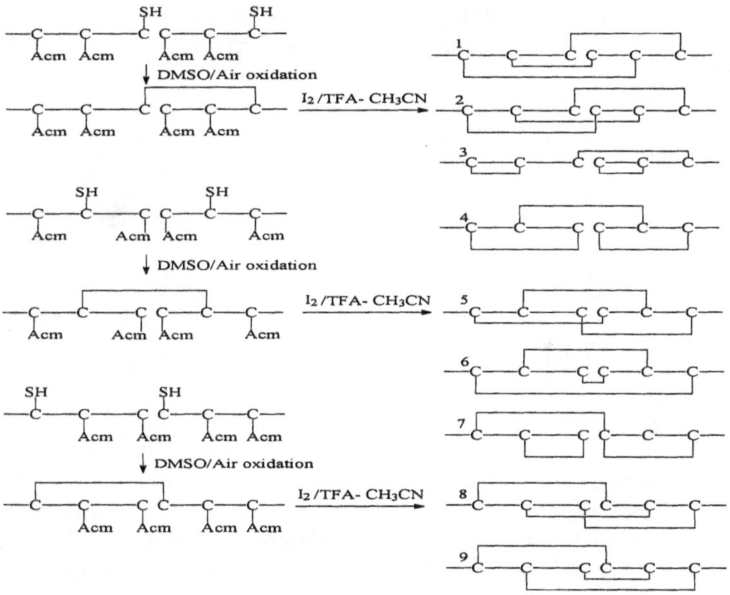

Fig. 2. Three logical oxidative folding approaches. Seven tricyclic peptides would be produced in theory, 2, 5, and 8 are the same.

To achieve the native-like peptide, three logical linear peptides were synthesized, each with Mbzl blocking one of the three pairs of cysteines involved in disulfides: Cys^{3} to Cys^{18} (SS^{1-4}, 1#), Cys^{10} to Cys^{22} (SS^{2-5}, 2#), Cys^{17} to Cys^{27} (SS^{3-6}, 3#) (fig. 2). Three monocyclic peptides (single disulfide peptide) were oxidized under the same conditions, respectively. The oxidation product (tricyclic peptide) possessed the same retention time as would the native peptide when mixed and co-eluted on HPLC.

Characterization of linear and folded peptides

Products were isolated by preparative RP-HPLC. The crude profiles for each oxidation procedure are shown in Figure 3. The amino acid composition of the synthetic peptides was examined by amino acid analysis, and was found to agree with the sequence (data not shown). The molecular weights of the synthetic peptides were determined by MALDTOF-MS (matrix-assisted laser desorption mass spectrometry), FAB-MS, and ES-MS (electric spray mass spectrometry), and all had the correct molecular weight (table 1).

Disulfide mapping experiments were carried out at 5 °C (fig. 4). 1B, 2B and 3B (B-shows the oxidative peaks which have the same retention times on HPLC as the monocyclc peptide, respectively) were mixed and coeluted on HPLC at an iso-gradient of 38% CH_3CN, which indicated they were the same. The monocyclic peptide 3# produced a main tricyclic product —3B. The yield was up to 58.3%, showing that the disulfide bond $SS^{3\text{-}18}$ induced folding to form the native disulfide pattern very well. The yields of 1B and 2B were 31.2% and 19.7%, respectively, and the yield of 3B was up to 66.7% at 18 °C.

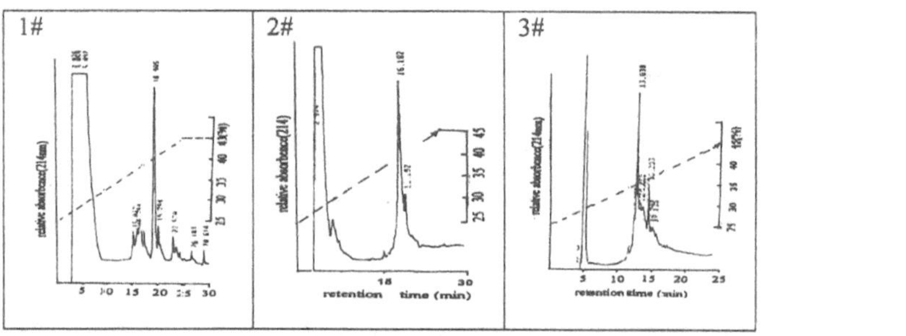

Fig. 3. Reversed-phase HPLC profiles of the linear peptides in the first step. Panel 1#, 2#, 3# show the results of three disulfide pattern. Peptides were eluted with a 25-45% buffer B linear gradient over 25 min at 1ml/min, monitored at 214 NM. Buffer A was 0.1% TFA, buffer B was acetonitrile contained 0.1% TFA. Cromasil C18 column (25 × 0.45 cm,5 μm particle size) was used.

Discussion

Our approach to form 1 disulfide in the first step and then 2 disulfides in the second step has been employed successfully to synthesize δ–conotoxin PVIA. There are three logical blocking modes corresponding to each of the three desired disulfide pairs. Three logical monocyclic peptides were oxidized at the same time and conditions. The oxidation product possessed the same retention time as would the native peptide when mixed and co-eluted on HPLC. The method is useful for finding the desired peptide without a standard sample.

Because the loop of Cys^3 and Cys^{18} is longer than that of the other two and the disulfide bond SS^{3-18} formed first could stabilize the secondary structure of the desired peptide, the yield of the main product is good. Our strategy is easy to employ for the synthesis of δ–conotoxin-like peptides.

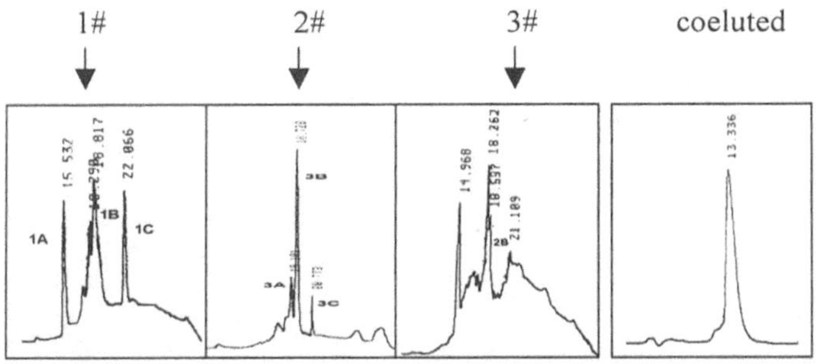

Fig. 4. Reversed-phase HPLC profiles showing the formation of tricyclic peptides. Peptides were eluted with the same buffer linear gradient as in fig.3. 1B, 2B, and 3B were mixed and coeluted on HPLC at an iso-gradient of 38% CH_3CN, which indicated they were the same. Cromasil C18 column (25 × 0.45 cm, 5 μm particle size) was used.

Table 1. Molecular weight determined by mass spectrum for peptides with or without disulfides.

Name	Theoretical M.W.	Determined M.W.
1# monocyclic	3286.9	3287.2 FAB
2# monocyclic	3286.9	3286.3 FAB
3# monocyclic	3286.9	3285.94 +/-0.41 ES
1#Tricyclic 1B	2998.6	2998.8 FAB
2#Tricyclic 2B	2998.6	2998.2 FAB
3#Tricyclic 3B	2998.6	2997.8 FAB
2# linear	3288.9	3289.9 MALDTOF

References

1. Möroder, L., Biopolymers (peptide Science), 40 (1991) 207.
2. (Jiang, H., Miao, Z.H., Zhong, M.N. and Chen, J.S., Youjihuaxue, unpublished.)
3. Dürienx, J.P., and Nyfeler, R., Peptides, 23 (1995) 165.
4. Fainzilber, M., Lodder, J.C., Kits, K.S., Kofman, O., Vinnitsky, I., Rietschoten, J.V., Zoltkin, E., and Cordon, D., J. Biol. Chem., 270 (1994) 1123; Fainzilber, M., Kofman, O., Zoltkin, E., and Cordon, D. , J. Biol. Chem., 269 (1994) 2574.
5. Shon, K. J., Grilley, M.M., Yoshikami, D., Hall, A.R., Kurz, B., Gray, W.R., Imperial, J.S., Hillyard, D.R., and Olivera, B.M., Biochemistry, 34 (1995) 4913.
6. Tam, J.P., Wu, C.R., Liu, W. and Zhang. J.W., J. Am. Chem. Soc. 113 (1991) 6657.

Session B
Structure, folding and conformation analysis
De novo design of peptide & protein

Chairs: Victor Hruby
University of Arizona
Tucson, U.S.A.

Gui-Shen Lu
Institute of Materia Medica CAMS
Beijing, China

Design, synthesis and NMR structure of a rigid cyclic peptide template

Steven A. Muhle and James P. Tam
Department of Biochemistry, Vanderbilt University, Nashville, TN 37232, USA

Introduction

Peptide dendrimers and branched peptides are designed artificial proteins with peptides centrally attached to a template or a polyamino-acid dendron [1-4]. Generally, 2 to 16 peptidyl branches of the same or different sequences are found in this design of multi-chain, cascade-shaped biopolymers. Our laboratory has previously developed and extensively exploited the Lys_2Lys (K_2K) or di-K_2K dendrons as templates to tether four or more units of peptides [4]. These types of flexible polyamino-acid templates are used in the design of multiple antigenic peptides tools for biological research [1]. Others [2-4] have devised constrained templates, often derived from non-peptide sources, to form branched peptides with four or less units with the intention that such templates would assist the folding process to form artificial proteins. Examples include cyclic peptides as templates developed by Mutter and his associates' [2] to provide stability and control in the orientation and positioning of the tethering peptides to a desired tertiary structure. Similarly, a variety of rigid templates have been popular in the design of artificial proteins based on the four-helix bundle structures [2-6].

Recently, Wong, *et al.* [4] have shown that the primary factor to stabilize a four-helix bundle structure of a branched peptide is due to the intra-molecular associations as a result of the covalently linkage of peptides bundled as a dendrimer. They conclude that the type of template used in four-helix bundle formation is not a determining factor to induce the helical structure. Since we have been interested in studying the properties of different peptide templates, the findings by Wong *et al.* have prompted us to compare the flexible K_2K dendron with a very rigid peptide template to influence various structural motifs. We envisioned that a cyclic peptide template, which is cystine-stabilized, anti-parallel β-sheet and (fig. 1) would provide the rigidity in comparison with the tetravalent K_2K dendron template in the formation, stability, and function of various structural motifs. Our template design is a 16-amino-acid cyclic peptide with two Pro-Gly turns. Two cross-bracing disulfide linkages provide further constraint to position four Lys (or other functional groups) on one surface in a unidirectional format to tether peptides.

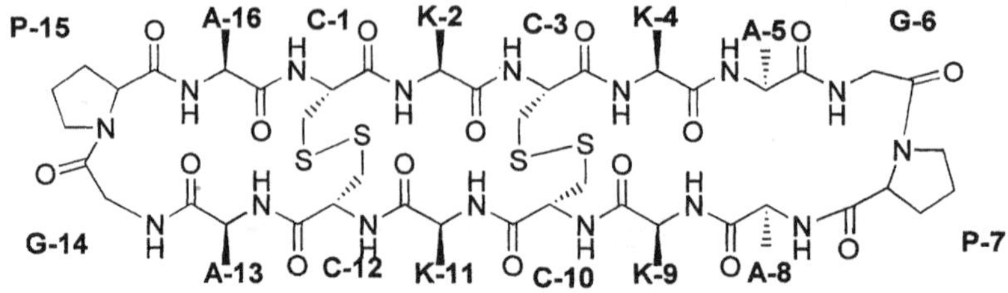

Fig. 1. Proposed template structure and amino acid sequence.

This paper describes the synthesis and structural characterization by 2D NMR of the cyclic rigid template. The results based on chemical characterization by MALDI-MS and NMR experiments by TOCSY, COSY, and NOESY are consistent with our intended design of a rigid cyclic peptide template with anti-parallel β-sheet structure, type II β-turns, and the desired cysteine connectivity, as shown in Fig. 1.

Results and Discussion

Our synthetic plan consisted of two tandem reactions. An on-resin stepwise solid phase synthesis by Boc-chemistry was used to prepare a linear peptide precursor as a C-terminal thioester. Following the removal of all protecting groups such as Lys(ClZ) adn Cys(MBzl) by HF, an off-resin cyclization of the crude and free linear peptide in an aqueous condition was used to form the cyclized template. Because of the circular permutated nature of the template, cyclization could be performed at any one of the four Cys-Xaa bonds, the least hindered Cys-Ala bond was chosen and the linear sequence of CKCKAGPAKCKCAGPA was used for the stepwise synthesis. The C-terminal thioester was prepared from a *p*-methylbenzhydrylamine resin coupled to mercaptopropionic acid to give a removable thiopropionyl linker. Coupling of Boc-Ala by the Bop activation formed the C-terminal Boc-Ala-thioester resin [6].

Cyclization by the orthogonal thioester method [6] between the two termini to form a Cys-Ala bond was initiated by one of the four cysteinyl thiols, acting as a nucleophile that attacks the weakly activated C-terminal thioester to form a thiolactone under an aqueous condition at pH 7.6. This transthioesterification is under the control of ring-chain tautomerization. Since there are four thiols present in the sequence, a series of thiol-thiolactone exchange referred to as a thia zip reaction by our laboratory, facilitates the formation of an N-terminal thiolactone through smaller thiolactone intermediates. Once the N-terminal thiolactone is formed, a spontaneous ring contraction via an S, N acyl shift affords the desired Cys-Ala lactam. The resulting cyclic peptide was immediately oxidized with 10% DMSO in a buffered aqueous solution at pH 7.6 and purified by C18 HPLC. The purified product determined by MALDI-MS had a mass of 1,514 daltons, which agrees the calculated mass. No oligomerized or end-to-side chain products were detected as monitored by HPLC and MALDI-MS analyses. Thus, our synthetic scheme for preparing rigid cyclic peptides is facile, efficient, and more importantly, regiosepecific. It offers significant improvement over the conventional methods of cyclization using protected peptide segments and enthalpic activation.

1D and 2D NMR determined structural characterization. The position of the disulfide bonds was determined by comparison with a 1D spectrum of a similar cyclic peptide differing only with α-aminobutyric acid substituting for Cys 1 and 12 (data not shown). The αH chemical shift for Cys 3 and 10 (4.70, 4.63), present in both peptides, is identical indicating that the same disulfide bond is present in both peptides. The 2D double quantum filtration correlated spectroscopy (DQF-COSY), total correlated spectroscopy (TOCSY), and nuclear Overhauser effect spectroscopy (NOESY) experiments were performed under aqueous conditions on a 500 Mhz Bruker NMR using Watergate graduated water suppression. The residue assignments and chemical shifts are presented in Table 1. Spin systems were identified on the TOCSY spectra, and proton assignments were made according to the sequential NOESY connectivities. The ^{3}JN, α vicinal couplings were measured from the DFQ-COSY spectra.

The dipolar connectivity's obtained from the NOESY experiments (fig. 2) were used to determine secondary structural elements. Strong sequential and long-range NOE crosspeaks were present to suggest an anti-parallel β sheet structure.

Table 1. Chemical shifts in 5% D_2O/H_2O at 25°C, pH 5.0.

	NH	Hα	Hβ	Hγ	Hδ	Hε
C-1	7.53	4.48	2.85,2.96			
K-2	8.66	4.38	1.74	1.30	1.62	2.92
C-3	8.19	4.70	2.96,3.04			
K-4	8.68	4.24	1.74	1.37	1.62	2.92
A-5	8.04	4.26	1.34			
G-6	7.93	3.83,4.34				
P-7		4.32	2.20	1.92	3.51,3.67	
A-8	8.08	4.22	1.32			
K-9	7.98	4.20	1.74	1.34	1.62	2.92
C-10	8.25	4.63	3.03,3.08			
K-11	8.71	4.38	1.83	1.30	1.63	2.92
C-12	7.77	4.48	2.88,3.21			
A-13	8.51	4.48	1.30			
G-14	8.28	4.36,3.88				
P-15		4.32	2.20	1.92	3.51,3.67	
A-16	8.38	4.32	1.29			

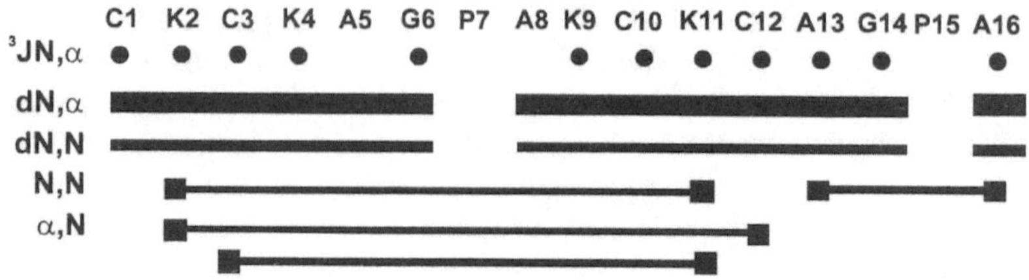

Fig. 2. Observed ^{3}JN, α coupling and summary of NOE data. (Black dots indicate coupling greater than 8 Hz. An open rectangle indicates a strong NOE; a black line is a weaker NOE. Open squares connected by a line are long range NOEs.)

Values of ^{3}JN, α greater than 8 Hz, such as those found in all lysines and cysteines, and in two of the alanines, are characteristic of a stable extended, β-sheet conformation. Alanine 5 and 8, which both couple at 6.4 Hz, are suggestive of a bulge or bend at one end of the peptide. This may be due to the C-3-->C-10 turn being larger, and therefore, more flexible than the C-12 --> C-1 turn. The strong dN, N connectivity's with weaker dN, N connectivities are also characteristic of an anti-parallel β-sheet structure [7]. This conformation is further supported by long range NOEs between the amides of A-13 --> A-16, K2 --> K-11, and the Hα/amide interaction between C12 --> K2, C3 --> K-11. These residues must be in relatively close proximity to generate the NOE to account for an anti-parallel β-sheet conformation. The strong coupling of the glycine residues and the innate structural constraints of the prolines suggest β type II turns. The identified structure places all four lysines in a unidirectional configuration.

In summary, the structural data is consistent with a rigid anti-parallel β-sheet template design. This template is similar to the RAFT templates which Dumy, *et al.* has prepared and characterized [8]. It is important to note that our structure is determined in aqueous conditions, which may be more relevant than the previous work [8], which are determined in organic solvents. The addition of the two-disulfide linkages should also provide greater rigidity to the template. Such a rigid template design will be useful for a comparative study with the more flexible K_2K dendron template to determine their effect to influence tertiary structures as a result of dendrimer formation and in the design of *de novo* peptides and proteins.

Acknowledgments

This work was in part supported by U.S. Public Health Service NIH grants AI46164 and CA36544, and also by the Medical Scientist Training Program at Vanderbilt University.

References

1. Tam, J. P. Proc. Natl. Acad. Sci. 85 (1988) 5409.
2. Tuchscherer, G. and Mutter, M. J. Biotechnology, 41 (1995) 197.
3. Schneider, J. and Kelly, J.. Chem. Review, 96 (1995) 2169.
4. Wong, A. et al. J. Am. Chem. Soc., 120 (1998) 3836.
5. Shao, J. and Tam, J.P. J. Am. Chem. Soc., 117 (1995) 3893.
6. Zhang, L. and Tam, J.P. J. Am. Chem. Soc., 119 (1997) 2363.
7. uthrich, K. NMR of Proteins and Nucleic Acids. Wiley-Interscience, New York,
8. (1986).
9. Dumy, P. Biopolymers, 39 (1996) 297.

Automated docking of Sch68631 with HCV NS3 protease

Xiao-Jie Xu[a], Ting-Jun Hou[a], Li-Rong Chen[b], Zheng Li[a] and Jia-Quan Wang[a]

[a]Chemistry Department, Beida-Jiuyuan Molecular Design Laboratory, Peking University, Beijing, 100871; [b]Department of Technique-Physics, Peking University, Beijing, 100871, China

Introduction

Hepatitis C virus has been identified as the major causative agent for most cases of non-A, non-B heptatis [1]. It may establish a chronic infection persisting for decades, usually resulting in recurrent and progressively worsening liver inflammation, that often leads to cirrhosis and hepatocellular carcinoma. The HCV NS3 serine protease plays the major role in mediating all the cleavages of the HCV polyprotein at the junction subsequent to NS3 in the nonstructure segment. The crystal structures of NS3 and NS3-NS4A complex were resolved by Love *et al* [2] and Kim *et al* [3], respectively. This information derived from the protease will stimulate efforts to develop more effective anti-HCV therapies. In previous years, several types of compounds were identified as HCV proteinase inhibitors. Among them, Sch68631, a new compound from the fermentation culture broth of *streptomyces* sp. [4], was an effective inhibitor. Until now, however the crystal structure of the complex was not obtained. It is therefore necessary to identify the binding mode, and thus to help in the design of new potential inhibitors.

As is well known, molecular docking can be used to clarify the mechanism of molecular recognition between the ligand and its receptor. Some theoretic docking methods have been used to study the interactions of molecular recognition. Among them, the docking method proposed by A. U. Brock et al. is an automated procedure based on molecular dynamics [5]. Compared with other docking procedures, it allows the ligand and the receptor to be flexible during simulations, so that the ligand and the receptor can adjust their conformations to find their best binding orientation. We have applied this method to find the best binding mode between Sch68631 and the NS3-NS4A protease. We hope to elucidate this complex structure and by indicating the most important interactions affecting binding affinity, to aid in the design of new potential inhibitors.

Results and Discussion

The receptor model used in this paper has been resolved by Kim *et al.* This model is a complex of NS3 and NS4A. It has been shown that the central region of NS4A, residue 21 to residue 34, suffices to mimic the effects on the full-length protein and can produce relatively influential effects on the structure and function of the NS3. Thus, the NS3 protease complex with the NS4A cofactor is believed to play a central role in the replication and maturation of HCV. From experiments, the catalytic site has been determined to be His-57, Asp-81, and Ser-139 sited in a crevice.

The compound of Sch68631 is a new HCV inhibitor isolated from the fermentation culture broth of *streptomyces* sp.. The structure was elucidated by analysis of spectrosopic data and shown to be a new member of the phenantrenequinone family of compounds (See fig. 1). Sch68631 can produce a relatively strong interaction with NS3 protease. It is thus an optimal probe molecule to study the possible binding mode of NS3 protease with its inhibitors.

The crystal structure of the NS3-NS4A complex is available from the Brookhaven Protein DataBank. Hydrogens were added to the complex using the molecular design software InsightII, with titrating residues assigned their normal protonating states at neutral PH and partial charges assigned to every atom. The complex was minimized using the cvff force field to remove any steric overlap. The deviation of the nonhydrogen atoms from the crystal structure was less than 0.8 Å.

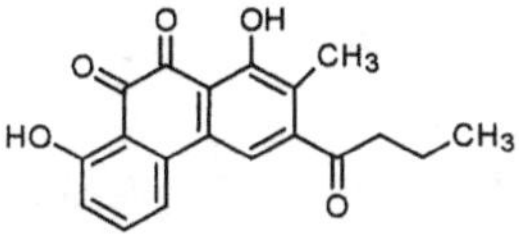

Fig. 1. Structure of Sch 68631.

The active site of NS3-NS4A has been identified by experiments [6] as the proposed catalytic triad of His-57, Asp-81, and Ser-139 residues. This catalytic triad is an ideal starting point for molecular docking to determine the binding mode. In this study, the mass center of the active site was chosen as the center of the subset for the defined binding site of the receptor. All the residues within 7 Å of the mass center of the active site are included in the subset. The remaining bulk atoms were assumed to be rigid. In computing the non-bonded interaction between the inhibitor and its protease, a cutoff of 15 Å was used. The compound Sch68631 was randomly placed near the active site as the initial position.

The NTV molecular dynamics method was applied to search the conformation space available to the system. A production run of 100000 steps with a time step of 1 fs (1.0×10^{-15} s) was performed and a constant temperature (500 K) was maintained during the simulation. The energy and coordinates were saved every 500 steps. The trajectory file from the molecular dynamics was used to cluster the conformations into groups. From every group, one conformer was randomly selected and minimized.

After calculation, cluster analysis was performed for the conformations in the trajectory file. An arbitrary criterion that a member of a group has a rmsd less than 0.5 Å from another member of that group results in 11 clusters. From every cluster, a conformer was randomly selected, yielding 11 representative conformers. All 11 conformers were fully minimized by using molecular mechanics with cvff force field.

The conformer with the lowest energy was treated as the best candidate of the binding mode. Fig. 1 shows the complex structure with the lowest non-bonded energy.

It seems that the probe molecule does not vary much in its initial structure. It basically resides near the active site, but calculation results also suggest that the binding site and the active site of NS3 protease is not completely the same. Of the three residues of the active site, only one residue, Asp 81, can form a hydrogen bond with sch68631. The other two residues seem not to generate direct interactions with sch68631. From the chemical environment around Sch68631, the best binding mode that is very energetically favorable consists of four hydrogen bonds between sch68631 and Ser42, Asp81, Lys136 and Gly137 of NS3 protease. Because the complex of NS3-NS4A is a relatively symmetric dimmer structure, another binding site must exist on the other domain of this complex.

Acknowledgments

This work is supported by NCSF and National Education Committee Foundation.

References

1. Choo, Q., Kuo, G., Weiner, A.J., Overby, L.R., Bradley, D.W. and Houghton, M., Science, 244 (1989) 359.
2. Kim, J.L., Morgenstern, K.A. and Lin, C., Cell, 87 (1996) 343.
3. Love, R.A., Parge, H.E. and Wichershan, J.A; Cell, 87 (1996) 331.
4. Chu, M. and Mierzwa, R., Tetrahedron Letters, 37 (1996) 7229.
5. Luty, B.A. and Wasserman, Z.R., J. Comp. Chem., 16 (1995) 454.
6. Hijikata, M. and Mizushima, H., J. virol., 67 (1993) 10773.

Molecular modeling of the three-dimensional structure of the human interleukin-11

Li-Rong Chen[b], Ting-Jun Hou[a], Xiao-Jie Xu[a*], Zheng Li[a] and Jia-Quan Wang[a]

[a]Chemistry Department, Beida-Jiuyuan Molecular Design Laboratory, Peking University, Beijing, 100871; [b]Department of Technique-Physics, Peking University, Beijing, 100871, China

Introduction

Interleukin-11 (IL-11) is a multifunctional cytokine possessing wide biological activities which can affect growth differentiation in several hematopoietic cell types, including early pluripotent stem cells, megakaryocyte progenitors and megakaryocytes, erythrocyte progenitors, etc. [1]. It is also identified as inhibiting adipogenesis in preadipocyces and stimulating production of several acute phase plasma proteins in hepatocytes. IL-11 has been shown to share many biological activities with IL-6, including the ability to activate expression of the same early response genes.

Until now, the 3D-structure of IL-11 has not been available from experiments, although large studies have been performed. From previous studies, it has been revealed that the cytokine family shares a common structural motif consisting of four anti-parallel helices in an up-up-down-down configuration. To date, the four-helix bundle fold has been observed in human growth hormone (HGH), interleukin-2 (IL-2), interleukin-4 (IL-4), granulocytecolony-stimulating factor (G-CSF), inter-leukin-5 (IL-5) and human macrophage colony-stimulating factor (GM-CSF). The cytokine family is one of the most diverse protein families but their structural relationships can be determined by comparing their amino acid sequences. From sequence alignment, it can be found that only G-CSF shares sequence similarity higher than 30% with IL-11.

In this paper, we have combined sequence comparison and secondary structural prediction to generate a molecular model of interleukin-11. The X-ray structure of G-CSF [2] was used as a template to produce the model, moreover, in the sequence alignment, the information from the secondary structural prediction of the GOR_II method [3] was taken into account. The validation of the model suggests that the quality of this structure is sound. The predicted model may aid in the design of experiments to better understand IL-11 and its interaction with receptors.

Results and Discussion

A pair-wise sequence alignment of IL-11 and G-CSF was produced using the Needleman and Wunsch algorithm [4]. When performing the sequence alignment, the penalty of a gap was defined to be proportional to the length of the gap. GOR_II method was applied to predict the secondary structure of IL-11. The sequence alignment of IL-11 and G-CSF was modified manually to adjust the structurally significant region of IL-11 according to information from the secondary structure prediction. Gaps were avoided in the predicted continuous helical regions. For model building of IL-11, the coordinates of the sequences of conserved regions came directly from the template protein. The coordinates for loops were assigned by searching the Brookhaven Protein Databank (PDB). In these calculations, a standard rotamer library was used and a cutoff distance of 10Å was applied for treatment of nonbonded interaction. In addition, the initial building model was energy minimized with CVFF force field until the root square mean (rms) derivative of the energy was less than 0.1 kcal/ (mol. Å). The completed model was validated using the Health program in Quanta and was checked from several aspects. Energy minimization calculations were carried out using the DISCOVER program in INSIGHTII (MSI, San Diego, CA, U.S.A), and the prediction of the IL-11 model was performed using the Homology program in INSIGHTII.

From the sequence database search we found that a total of three proteins shared a similarity larger than 30% with IL-11. But from the folding check of their 3-D structures, only G-CSF belongs to the same family as IL-11. G-CSF not only shares relative high sequence similarity with IL-11, but also possesses the same structural motif. Thus the X-ray structure of G-CSF was used as a reference protein to model the 3D-structure of IL-11. GOR-II method was carried out to predict the secondary structure of IL-11, but the contributions of the helical region had to be considered. The alignment was carefully modified manually so that the gaps appearing in the continuous helix were as few as possible. Fig. 1 shows the sequence alignment of G-CSF and IL-11 after manual adjustment.

Based on the alignment, the three-dimension model was constructed using the X-ray structure of G-CSF as the template. The coordinates of the sequences of the reserved regions were obtained from G-CSF except for those residues from 164-178. The analysis from the sequence alignment and secondary structure prediction suggested residues 149-174 should be a continuous helix, but sequence alignment introduced a gap between them that separates the helix into two parts. It is thus not appropriate to copy coordinates from G-CSF for this segment. For these residues from 164 to 178, the 3-D structures instead are modeled according to the standard structure of an α-helix. Intermolecular contacts and stereochemistry of the initial assembled model was refined by energy minimization using DISCOVER program.

	1	10	20	30	40 50
1GNC:	MTPLGP—A	SSLPQSFLLK	CLEQVRKIQG	DGAALQEKLC	ATYKLCHPEE
IL11:	Pgpppgpprv	spdpraelds	Tvlltrsl*la*	*dtrqlaaqlr*	*d*kf----pad
1GNC:	LVLLGHSLGI	PWAP-LSSCP	SQALQLAGC	SQLHSGLFLY	QGLLQAL—
			L		
IL11:	g---dhnlds	lptlamsaga	Lgalqlpgvl	trlradllsy	lrhvqwlrra
1GNC:	EGISPE-LGP	TLDTLQLDVA	DFATTIWQQ	EELGMAPALQ	
			M		PTQGAMPAFA
IL11:	Ggsslktlep	elgtlqarld	Rllrrlqllm	srlal-pqpp	pd---ppapp
1GNC:	SAFQRRA-G	VLVASHLQSF	LEVSY-RVLR	HLAQP	
IL11:	Lappssawgg	*raahailgg*	*L*hltl*dwavr*	*gllll*ktrl	

Fig. 1. Sequence alignment of the G-SCF and IL-11. (A) The gap appearing in the alignment is labelled with -. (B). The italic letters indicate the helices predicted using GOR_II method. (C). The PDB code for G-CSF is indicated with 1GNC. (D) The G-CSF sequence which already has 3D-structure is denoted as upper letters, the IL-11 sequence without 3D-structure is denoted as lower letters.

From the validation of the model, we obtained some interesting information. The structure did not have any close contact atoms. The main chain conformation agreed with generally accepted regions on the Ramchandran map. Moreover, the exposed hydrophobic side chain only contained several residues on the N termini, and the buried hydrophilic side chain was composed of only residues 91 to 102. A qualitative judgement indicated the model generally was rational from the perspective of the structure and chemical environment of the protein.

In the model (fig. 2), four α-helices were connected by peptide segments of variable length that could be differentiated in the proposed structure. Such an arrangement of helices and connecting segments is typical of a four-helix bundle topology, a motif which possesses the distinct characteristics of helical cytokine family (fig. 3). The preservation of the four-helix bundle topology in the members of the cytokine family may be important for their interaction with their receptors. The predicted model provides a first approximate view of the IL-11 structure and gives some important information for further study of IL-11 itself and its interaction with its receptors.

Acknowledgments

This work is supported by NCSF and National Education Committee Foundation.

References

1. Paul, S. R.and Schendel. P., Int. J. Cell Cloning, 10 (1992) 135.
2. Hill, C. P., Osslund, T. D.and Eisenberg, D., Proc. Natl. Acad. Sci., 90 (1993) 516.
3. Garnier, J.and Robson, B., Predictions of Protein Structure and the Principles of Protein Conformation, New York, Ch., 10 (1989) 417.
4. Needleman, S.B.and Wansch, C. D., J. Mol. Biol., 48 (1970) 443.

Designing β-hairpin forming short peptide

Yin-Lin Sha, Yong-Liang Huang, Qi Wang and Lu-Hua Lai
Institute of Physical Chemistry, Peking University, Beijing, 100871, China

Introduction

α-helix and β-sheet are the most abundant elements in native proteins. In addition to α-helix, β-sheet has also been suggested as the possible site for nucleation in protein folding [1]. Because of β-sheet aggregation, it is very difficult to elucidate the folding mechanism [2]. β-hairpin, the super-secondary structure composed of two strands connected with a β-turn, was selected as a template for studying the interactions of β-sheets. In the past few years, studies have mainly focused on β-hairpin peptide from natural proteins, and β-hairpins in isolated protein fragments or modified ones were reported to be stable in water or organic solvents [3-5]. The shortest β-hairpin peptide, an eight residue linear moieties of de novo design in crystal form was reported by Satish K. Awasthi *et al* [6].

According to reports [3-5], at least three main factors are important in β-hairpin formation: main chain interaction, hydrophobic interaction and the β-turn type. Significantly, the solvent effect on β-hairpin formation was discussed in most recent reports [3-5]. We have designed a short β-hairpin forming peptide and studied its solution conformation by CD spectra.

Peptide design

A short peptide containing eight amino acid residues was designed to form a β-hairpin. The type II' β-turn D-ProGly reported by Struthers et. al. [7] was used to enhance its stability, and residues $Val_{1,6,8}$ and Leu_3 with high β-structure propensity were utilized to construct the two arms. $Threonine_{2,\ 7}$ with γ-methyl and γ-hydroxyl were used to enhance the interactions between the arms.

Material and methods

The peptide H_2N-Val-Thr-Leu-D-Pro-Gly-Val-Thr-Val-COOH was synthesized by solid phase peptide synthesis method with HBTU/HOBT strategy [8]. The first amino acid was loaded to Wang resin (0.7 mmol/g, substitution) by amino acid anhydride strategy. The Fmoc protected amino acids FmocVal, FmocLeu, Fmoc-D-Pro, FmocGly and FmocThr (But) (ACT, USA) were used in the synthesis. The peptide resin was cleaved by the K reagent (TFA: H_2O: Phenol: Thioanisole: EDT 82.5:5:5:5:2.5).

The crude product was purified by RP-HPLC (Gilson, France) with C_{18} Cosmosil column. The flow phase was water/acetonitrile (containing 0.1% TFA) system. The zorbax RP-C_{18} (4.6 × 250 mm) column was used for peptide analysis. Molecular weight was confirmed by mass spectra.

CD spectra were recorded on Jobin Yvon CD6 with 0.1 mm cylinder cell, 186-260 nm, 20 °C. The peptide solution was prepared using phosphate (pH 7.04) buffer.

Results

Fig. 1 shows the HPLC profile of the peptide.

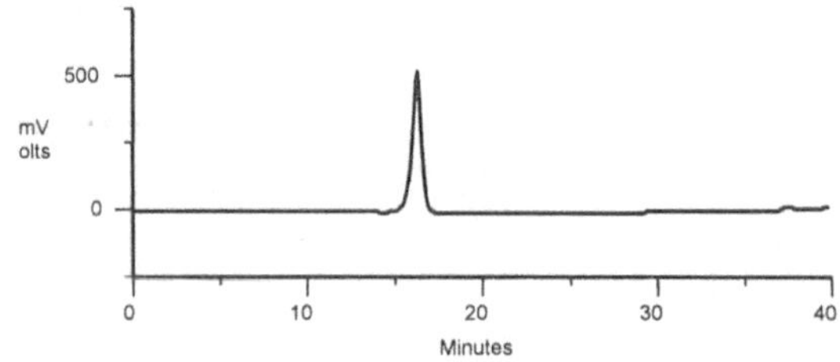

Fig. 1. The HPLC analytic spectra of the designed peptide. Zorbax C18 column 4.6 × 250 mm, flow rate 1 ml/min, flow phase: acetonitrile/water.

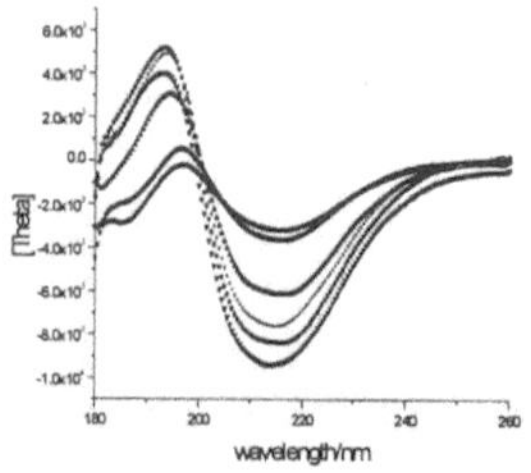

Fig. 2. Far-UV CD spectra of the designed peptide under different TFE concentrations. From less to more negative [theta] at 216 nm is 0.0%-50% TFE.

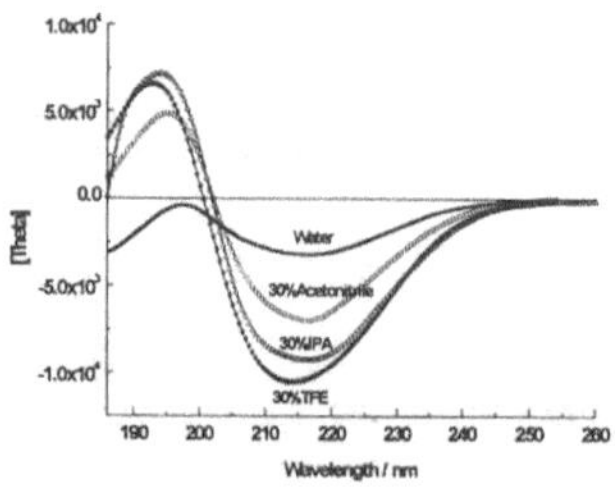

Fig. 3. Far-UV CD spectra of the designed peptide in PBS buffer containing different organic solvent (30% acetonitrile, 30% isopropyl alcohol or 30% TFE)

Fig. 2. shows the CD spectrum of the designed peptide. The spectrum has a maximum at ca. 194 nm and a minimum at ca. 216 nm, a typical β-hairpin characteristic [9]. Johnson postulated such parameters could be expected for a β-sheet mixed with a β-turn [10]. In TFE titration, the spectrum varies with TFE concentration (10%~50%) and shows an isodichroic point at ca. 200 nm which indicates that the folded structure in water has been stabilized.

Fig. 3 is the CD spectrum of the designed peptide in PBS buffer containing 30% acetonitrile, 30% isopropyl alcohol or 30% TFE (volume ratio). The peptide behaves similarly in these organic reagents. The maximum and minimum of the spectra are enhanced with TFE, isopropyl alcohol or acetonitrile titration. At the same time, a similar isodichroic point (at ca. 205 nm) of the spectra is observed with titration by different organic reagents.

Discussion

β-turn is one of the most important factors in determining the stability of β-hairpins. We have reported that D-ProSer behaves better than GlySer [11]. Haque T.S. *et al* also reported that the hairpin could be stabilized with enforced turns like D-ProD-Ala, D-ProGly or D-ProAla [12]. They also pointed out that the conformational proclivity of the backbone is at least as important as hydrogen bonding and hydrophobic interactions in stabilizing β-hairpin conformations. We think that β-turn is particularly more important in a short β-hairpin peptide than in a protein or a long peptide.

L. Serrano *et al* [9] have reported a β-hairpin peptide BH8 whose CD spectrum in TFE 30% has a maximum at 202 nm and a minimum at 216 nm. Similarly, the peptide we report here also has a maximum at 194 nm and a minimum at 216 nm of its CD spectrum (fig. 1). This kind of spectrum characteristics has been postulated as what could be expected for a β-sheet mixed with a β-turn. With TFE titration, an isodichroic point (at ca. 200 nm) also appeared in the CD spectrum of the designed peptide, which is indicative of a two-state transition and means the structure in water has been stabilized [10].

The mechanism of interaction between TFE and α-helix has been discussed by Luis Serrano *et al* [13]. Compared to water, TFE does not destroy the hydrogen bonding of helix stabilization, but supports a good hydrophobic environment around the helix hydrophobic surface. In this work, the designed β-hairpin peptide also has a hydrophobic side. We postulate that the hydrophobic side tends to be more regular in the presence of TFE, which interacts with the peptide via its hydrophobic group. To test this, we have studied the conformational properties of the peptide with different organic solvent titrations. Similar to TFE, organic reagents such as acetonitrile and isopropyl alcohol can enhance β-hairpin stabilization. This demonstrates that TFE, isopropyl alcohol and acetonitrile obey a similar mechanism of β-hairpin induction. These three organic solvents are similar in the sense that they all have a hydrophobic group connected to a polar group. Thus we propose that hydrophobic interactions between peptide and organic solvent also play an essential role in β-hairpin stabilization. To support this opinion, we calculated the logP of TFE, isopropyl alcohol and acetonitrile (the values are + 0.41, + 0.05 and - 0.34), which show a good correlation with their β-hairpin inducing ability. In an amphilic environment, the more hydrophobic the potential solution is, the more stable the designed β-hairpin peptide.

We have designed an eight-residue β-hairpin forming peptide. CD spectra show the peptide forms a β-hairpin like structure, which can be enhanced by organic solvent.

From this study, we can conclude that a strong β-turn and amphiphilic environment are key factors for β-hairpin formation and stabilization.

Acknowledgments

This work was supported by the Department of Science and Technology of China and the National Natural Science Foundation of China. Thanks Mr. Ren-Xiao Wang for his calculating logP and valuable discussing.

References

1. Blanco, F. J.and Jimenez, M.A., Biochemistry, 33 (1994) 6004.
2. Osterman, D. G.and Mora, R., J. Am. Chem. Soc., 106 (1984) 6845.
3. Malashkevich, V.N. and Kammerer, R.A., Science, 274 (1996) 761.
4. Nautiyal, S. and Woolfson D.N., Biochemistry, 34 (1995) 11645.
5. Hocks, M.R.and Holberton D.A., Folding and Design, 2 (1997) 149.
6. Awasthi, S. K., Biochem. Biophys. Res. Commun., 216 (1995) 375.
7. Struthers, M.D. and Cheng, R.P., Science, 271 (1996) 342.
8. Knorr, R. and Trzeciak, A., Tetrahedron Lett., 30 (1989) 7192.
9. Marina, R.A., Blanco, F.J.and Serrano, L., Nature Structural Biology, 3 (1996) 604.
10. Johnson Jr, W.J., Ann. Rev. Biophys. Biophys. Chem., 17 (1988) 145.
11. Cao, A.N. and Sha, Y.L., Protein and Peptide Letters, 5 (1998) 53.
12. Haque, T.S. and Gellman, S.H., J. Am. Chem. Soc., 119 (1997) 2303.
13. Arthur, C.G. and Allen, T.J., J. Am. Chem. Soc., 118 (1996) 3082.

Anti-tumor activities of *papavor somniferum* pollen peptides

Jia-Xi Xu* and Sheng Jin
Department of Chemistry, Peking University, Beijing, 100871; National Laboratory of Applied Organic Chemistry at Lanzhou University, Lanzhou, 730000, China

Introduction

Peptides from certain pollen have been widely studied as allergens in the past three decades [1-5]. Plant pollens are not only important causes of allergy, but also nutritious and healthful food in many countries [6]. In recent years, our working group has paid much attention to studies on active oligopeptides in some special pollens (rape and opium pollens) and has obtained several peptides [7-9]. Bioassay results showed that they have activity in restraining some cancer cell division.

Results and Discussion

A pollen peptide fraction was obtained from the water extract of *papaver somniferum* pollen after separation on a sulfonated ion exchange resin column and a Sephadex G25 gel filtration column. The peptide fraction of *papaver somniferum* pollen was further separated to give inpure pollen peptides PSPP2 and PSPP3 (*Papaver Somniferum* Pollen Peptide 2 and 3) on a Dowex 50w x 2 ion exchange column with a linear gradient eluent from pH 3.08 to 5.06. Pure peptides PSPP2 and PSPP3 were obtained after further purification on reverse phase HPLC [9, 10]. Their sequences are PSPP2 NQQPLQTSGVINMKAAG and PSPP3 NQNGSNPKTVKQA.

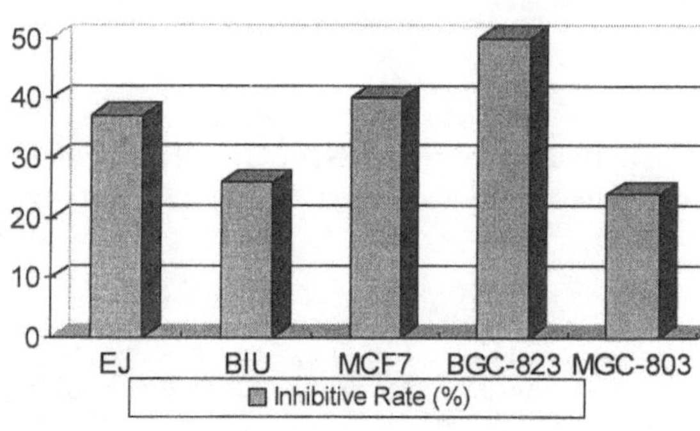

Fig. 1. Inhibitive rates of PSPP2 to tumour cell strains.

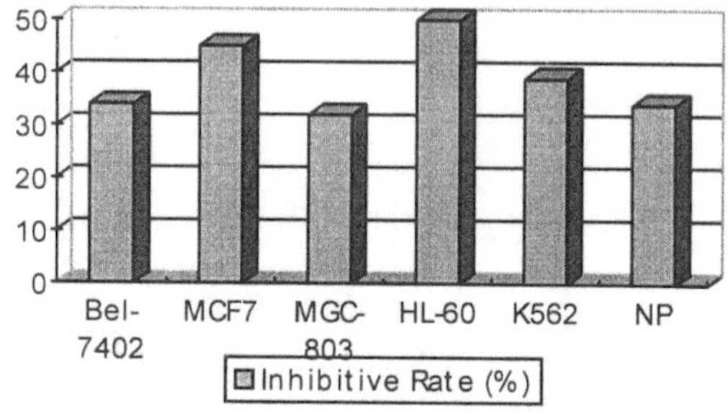

Fig. 2. Inhibitive rates of PSPP3 to tumour cell strains.

Peptide anti-tumor activities were evaluated by determining the percent inhibition against some tumor cell strains with the tetrazolium salt MTT [3-(4, 5-dimethylthiazol-2-yl)-2, 5-diphenyl tetrazolium bromide] method [11]. The tumor cell strains included human liver, Bel-7402 cells, gastric BGC-823 and MGC-803 cells, bladder EJ and BIU cells, mammary gland MCF7 cells, nasal and pharynx NP cells, leukemia HL-60 cells and red leukemia K562 cells.

The bioassay was carried out in half-area flat-bottomed microtiter trays at 10^{-4} mol/L of peptide. Each well contained the cells to be tested with the culture medium removed. Supernatant removal was accomplished by inverting, flicking and blotting the plate. If the cells had been grown in the well no prior centrifugation was needed; otherwise, supernatant removal was preceded by centrifugation. To the cells in each well was added 50 μL of a 1mg/mL solution of tetrazolium salt MTT in RPMI-PR^{-} medium. The tray was gently shaken and incubated for 3 h at 37 °C. At the end of the incubation period the plate was centrifuged and the untransformed MTT removed by carefully inverting, flicking and blotting the tray. 50 μL of propanol was added to each well. The plate was then vigorously shaken to ensure solubilization of the blue formazan. The optical density of each well was measured using an automatic plate reader with a 560 nm test wavelength and a 690 nm reference wavelength. The results are shown in Fig. 1 and Fig. 2.

Acknowledgments

Project (29572034) supported by National Natural Science Foundation of China and Bioorganic Molecular Engineering Laboratory at Peking University.

References

1. Suphioglu, C., Singh, M.B. and Knox, R.B., Int. Arch. Allergy Immunol., 102 (1993) 144.
2. Komiyama, N., Sone, T., Shimizu. K., Morikuba, K. and Kino, K., Biochem. Biophys. Res. Commun., 201 (1994) 1021.

3. Ansari, A.A., Shenbagamurthi, P. and Marsh, D.G., J. Bio. Chem., 264 (1989) 11181.
4. Ferreira, F.D., Hoffmann-Sommergruber, K., Breiteneder, H., Pettenburger, K., Ebner, C., Sommergruber, W., Steiner, R., Bohle, B., Sperr, W.R. and Valent, P., J. Bio. Chem., 268 (1993) 19574.
5. Batanero, E., Villalba, M. and Rodriguez, R., Mol. Immunol., 31 (1994) 31.
6. Wang, K.F., The Nutritive Value and Food Therapy of Pollen, Peking University Press, Beijing, 1986.
7. Jin, S., Xu, J.X. and Miao, P., Youji Huaxue, 13 (1993) 202.
8. Zhang, Y.L. and Jin, S., Chin. Chem. Letts., 1 (1990) 85.
9. Xu, J.X. and Jin, S., Chin. Chem. Letts., 4 (1993) 213.
10. Xu, J.X. and Jin, S., Acta Chim. Sin., 53 (1995) 822.
11. Denizot, R. and Lang, R., J. Immunol. Methods, 89 (1986) 271.

Structural investigation of two pollen peptides by 2 D homonuclear NMR spectroscopy

Yan-Ling Song[a], Jin-Feng Wang[b], Er-Cheng Li[b] and Ze-Chuan Qu[a]
[a]*College of Chemistry and Molecular Engineering, Peking University, Beijing, 100871 or National Laboratory of Applied Organic Chemistry at Lanzhou University, Lanzhou, 730000;*
[b]*Institute of Biophysics, Chinese Academy of Sciences, Beijing, 100101, China*

Introduction

We reported the peptide BPP1 extracted from buckwheat pollen at the last CPS in 1996. For structure-activity studies, an analogue of this peptide, BPP3, was synthesized by the solid phase method. Their sequences are as follows:

BPP1: A P V L Q I K K T G S N

BPP3: A P A L Q L K K N G S Q G NH_2

In the last decade, determinations of 3D solution structures of compact globular proteins and small inflexible cyclic peptides by NMR have been very successful. NMR also holds promise for providing structural information on flexible biomolecules, although a quantitative description of structure is complicated by the inevitable population-weighted averaging of the key NMR parameters. Here we report our NMR study on the two small peptides.

Methods

For water samples of BPP1 (10.2 mM, 90% H_2O/10% D_2O, pH=4.0), DQF-COSY, TOCSY (mixing time: 70 ms) and ROESY (mixing time: 256 ms) spectra were acquired in phase-sensitive mode on a Bruker DPX-400 NMR spectrometer at 26 °C. Other acquisition conditions are spectra width: 4000 Hz (10.0 ppm), data points: 2048; 128 scans for each of the 512, 512 and 420 tl increments for COSY, TOCSY and ROESY, respectively, and relaxation delay between scans: 2 s. For deuterium sample dissolved in 99.9% D_2O after three cycle of freeze-pump (10.0 mM, pH=4.0), DQF-COSY and TOCSY (mixing time: 65 ms) spectra were collected on a Bruker DMX-600 NMR spectrometer with the same acquisition condition as above, except for the spectra width, which is set at 3592 Hz (6.0 ppm) as all the amide protons disappeared after the freeze-pump treating of the sample. Data processing: window functions, sinebell, sinebell with 90 °C phase-shift and sinebell-square with 45 °C phase-shift, were applied to the COSY, TOCSY and ROESY spectra, respectively, and zero-filled to 2048 × 2048 data points before Fourier transformation.

Results and Discussion

The resonance assignment can be accomplished according to the standard procedure suggested by Wuthrich (table 1). As expected, there are 11 cross peaks in the finger-print region of the COSY spectrum of the water sample, which should result from the coupling between the amide protons and α protons of the 11 amino acids other than proline in BPP1. On the COSY and TOCSY spectra acquired in the water or deuterium solution, it is easy to find the signals from most of the protons in A1, P2, Q5, I6, T9, G10, S11 and N12, based on the characteristics of their spin systems. For example, the COSY spectrum of Gln-5 has cross peaks between its amide proton (8.40 ppm) and its α-proton (4.36 ppm), among its α-proton and β-protons (2.05, 1.95 ppm), among its β-protons and the γ-protons (2.33 ppm), respectively, and, on the TOCSY spectrum, all of these cross peaks are connected with each other. Although there is some signal overlapping to various extent for other spin systems, they can be unambiguously assigned with a combined use of COSY, TOCSY and ROEXY spectra. On the ROESY spectrum, the cross peak between the α-proton of P2 and amide-proton of V3 can be used to distinguish the spin-system of V3 from that of L4; the cross peak between α-proton of I6 and the amide-proton of K7 makes the spin-system of K7 differ from that of K8. The cross peak between the β-proton of T9 and amide-proton of G10 gives us further evidence for the assignment of the signals from the two α-protons of G10 with a small difference in chemical shift.

In terms of the chemical shift dispersion of amide- and α-protons, all the signals of BPP1 appear in narrow (for ranges amide-protons, 8.48~8.16 ppm; and for α-protons, 4.52~3.99 ppm), indicating that this peptide has no regular secondary structure. On its ROESY spectrum run with mixing times from 105 ms to 500 ms, all the assigned cross peaks are between the intra-residue or sequential proton pairs, and no cross peak between long range protons can be found. Using the assigned 35 ROE peaks as restraints, structure calculation with XPLOR program further shows that BPP1 has no regular conformation.

The spectra of BPP3 have an even more narrow range of chemical shifts than that of BPP1, which results in overlapping of the signals in the finger-print region. This implies that the average of the population-weighted conformations of this peptide in solution is different from that of BPP1. We will continue our investigation of this difference.

Table 1. The proton chemical shifts (ppm) of BPP-1.

	NH	α-H	β-H	Others
Ala-1	8.34	4.36	1.52 (CH_3)	—
Pro-2	—	4.52	2.32, 1.86	03 (γ-CH_2); 3.72, 3.61(δ-CH_2) (δ-CH_2)
Val-3	8.34	4.04	2.04	0.98, 0.93(γ-CH_3)
Leu-4	8.21	4.38	1.58	1.59(γ-CH_2); 0.93, 0.87(δ-CH_3)
Gln-5	8.40	4.36	2.05, 1.95	2.33(γ-CH_2)
Ile-6	8.21	4.13	1.83	1.48,1.18(γ-CH_2); 0.89, 0.86(γ,δ-CH_3)
Lys-7	8.41	4.34	1.76	2.98, 1.81, 1.67, 1.42
Lys-8	8.48	4.41	1.81	2.99, 1.86, 1.69, 1.41
Thr-9	8.2 9	4.36	4.23	1.2 1(γ-CH_3)
Gly-10	8.48	4.09, 3.99	—	—
Ser-11	8.23	4.51	3.86	—
Asn-12	8.16	4.5 1	2.29, 2.70	—

Acknowledgments

This work was supported by the National Natural Science Foundation of China (No. 29672002)

References

1. Rao, B.D.N. and Kemple, M.D., NMR as a structural tool for macromolecules. Plenum Press, New York, 1996.
2. Yao, J., Dyson, H.J. and Wright, P.E., J. Mol. Biol., 243 (1994) 754.
3. Wuthrich, K., NMR of Proteins and Nucleic Acids. Wiley, Chichester, 1986.

A comparison of three heuristic algorithms for molecular docking

Ting-Jun Hou[a], Xiao-Jie Xu[a] and Li-Rong Chen[b]
[a]Chemistry Department, Beida-Jiuyuan Molecular Design Laboratory, Peking University, Beijing, 100871; [b]Department of Technique-Physics, Peking University, Beijing, 100871, China

Introduction

The docking procedure between the inhibitor and the protein is a very sophisticated optimization problem. It is very difficult to carry out minimization using gradient methods such as the steepest descent method, the Gauss-Newton method, since it is too easy to fall into the local potential wells and very difficult to escape from them. Some heuristic methods therefore have been introduced into studies of molecular recognition. How to choose an adequate optimization method in the docking procedure is critical to the calculation of results.

In this paper, we describe the implementation and comparison of four search algorithms, random search (RS), simulated annealing (SA), genetic algorithm (GA), and Tabu search (TS), to the molecular docking procedure. The algorithms were compared using three protein-ligand systems.

Results and Discussion

In this study, the docking method that was applied to compare the algorithms is the two-stage soft-docking procedure developed by us [1, 2]. In our method, geometric complementarity and energetic complementarity are used as the score functions to evaluate the binding mode between the receptor and the ligand. Geometric complementarity is evaluated by the score of the matched surface dot areas minus the unfavorable atomic pairs. Energetic complementarity is evaluated by the non-bonded energy between the receptor and the ligand. Geometric complementarity alone can determine the proper binding mode between the receptor and the ligand to a bound complex; it has been proven that the minimum values of the geometric complementarity correspond to the preferred binding mode of the ligand [1, 2]. Moreover, our previous study showed that the score function of the geometric complementarity was smoother and simpler than the score function of the energetic complementarity. Thus using geometric complementarity better results may be obtained when these three heuristic algorithms are compared. In this study, therefore, geometric complementarity was used as the score function to evaluate the heuristic search algorithms.

Simulated annealing and genetic algorithms are two well-used heuristic algorithms which have been successfully applied in some docking procedures. More recently, Tabu search has begun to attract attention as an effective heuristic search procedure for combinatorial optimization problems in the molecular design field [3]. David et al were the first to apply this search method in docking procedures and proved it was very effective for finding the proper binding mode [4]. Because the docking procedure is a very complicated minimization problem, it is difficult to find an effective algorithm that can perform well in all conditions and the above three algorithms showed bad results in some cases. If using only a single algorithm, it is therefore very difficult to solve a docking problem thoroughly.

The first goal of this study was to compare these three algorithms, although comparison of algorithms is not easy. First, each category of algorithm has its own implementions, each of which will perform differently for a given optimization problem. Second, the performance of each algorithm depends on a set of adjustable operational parameters, and the quality of the results depends on whether they are optimal for a given test case. Thus in this study, in order to compare these three algorithms fairly, the methods are all implemented in a traditional manner that is not modified or revised. Moreover, the parameters applied in different algorithms are suitable in common cases.

Because all heuristic algorithms will contain stochastic elements, it is necessary to assess performance statistically over a sufficiently large number of independent trials. To ensure a fair comparison between algorithms, each one was limited to a maximum of 50000 ($\pm 1\%$) function evaluations per docking. This number is large enough for most algorithms to achieve convergence in most cases. When comparing the heuristic algorithms, the main quantity considered was the median score of the distribution of best scores obtained over 300 independent trials. In this study, the mean value of the maximum score of geometric complementarity was used as a descriptive statistical evaluation criterion to find the best docking mode. Moreover, we also compared algorithms according to their success rate proposed by Gehlhaar interquatile [5], the proportion of the trials which find a solution within 1.5 Å (heavy atoms only) of the crystallographic ligand conformation.

The comparison of algorithms was carried out over three test cases as specified in table 1. These three complex systems comprise two protein-small molecule complexes and one protein-protein complex. As the goal of this study was only to compare these three heuristic algorithms, the systems selected are all bound protein complexes. In order to simplify the calculations, the flexibility of the ligand and the receptor is ignored. The degrees of freedom are only six: three rotational degrees of freedom and three translational degrees of freedom of the ligand. Moreover, an active site for the receptor is defined in order to restrict the ligand docking in a small region. The active sites are all defined as a small box with $4 \times 4 \times 4$ Å in the calculations and the gravity center of the ligand must lie in this box.

Table 2 shows the test case results for the different algorithms. It is obvious that RS performs very poorly. The low mean score of the geometric complementarity and success rate reflect the fact that RS is ineffectual in a search space of this size. The success rate of 3DFR and 2WRP shows that GA performed best and TS performs a little worse. However the mean value of the 2WRP shows TS performs better than GA, so overall, TS performs a little better. The reason is that TS converges more slowly than GA, and in some cases, it may already reach near the best solution, but it cannot achieve the region that we have defined. Complex 2PTC is a protein-protein complex. From the mean value of the geometric complementarity and the success rate, the TS performs best. The contact surface between protein and protein is relatively large, so the score function of the geometric complementarity is more complex than that between protein and small molecule or protein and peptide. In this case, GA can fall into local minima easily and sometimes can not escape. TS, on the other hand, does better, since it can avoid falling into local minima.

Table 1. The test cases using in our calculation.

Molecular names	Probe atoms[a]	Targe atoms	Probe dots[b]	Target dots
3DFR	33	1343	128	2957
4MBN	44	1294	150	2871
2PTC	454	1629	1184	3347

[a] *The number of probe and targe atoms only represent the number of the heavy atoms.*
[b] *The probe radius to generate the connolly surface is defined as 1.5 Å*

From the comparison of these three algorithms, GA and TS both have their merits and shortages. GA converges faster. When near the best solution, it can find it very quickly, but can fall into local minima very easily. Contrasted with GA, TS can avoid falling into local minima, but it converges relatively slower. According to their merits and shortages, we thus proposed a hybrid algorithm (HA) that combined GA and TS. The basic procedure of the hybrid algorithm is similar to TS, but differs in two points. The first difference is that after N possible moves from the current solution, some extra steps of crossover and mutation operations, which come from GA, are added. The second modification is after every crossover operation, N new solutions are ranked, and the best several solutions compared with the solutions in the Tabu list to check if they are Tabu. If they are, these Tabu solutions are replaced by new solutions generated randomly. The new hybrid algorithm has the merits of GA and TS at the same time; it not only converges fast, but also does not easily fall into local minima. When we applied the hybrid algorithm to these three test cases, the calculated results showed this new algorithm did much better than the three heuristic algorithms alone.

Table 2. Docking results for the test cases given in table 1.

PDB code	Algorithm	Maximum score	rms/rmd(ligand)[a]	Mean score	Success rate(%)
3DFR	SA	1309.04	1.06	625.31	21
3DFR	GA	1340.37	1.01	701.51	33
3DFR	TS	1178.05	0.76	812.45	31
3DFR	RS	508.32	5.51	405.10	0
3DFR	HA	1268.11	1.12	1102.23	81
4MBN	GA	1230.45	0.77	724.50	51
4MBN	TS	1143.49	0.88	756.23	42
4MBN	RS	876.54	2.35	489.23	2
4MBN	HA	1223.23	0.68	1108.90	96
2PTC	SA	2177.14	0.70	1703.23	59
2PTC	GA	2184.62	0.78	1650.45	50
2PTC	TS	2194.56	1.07	1745.56	48
2PTC	RS	1935.87	1.89	1345.70	4
2PTC	HA	2209.67	1.09	1956.98	100

[a] *rmd means root mean distance, All the distances of atomic pairs of the ligand molecule have been calculated. The root mean of the different of relative distances (rmd) of bound complexes is used to evaluate the affectivity of the soft-docking calculations.*

Acknowledgments

This work is supported by NCSF and National Education Committee Foundation.

References

1. Wang, J.M., Xu, X.J. and Jiang, F.,In Xu,X.J.,Ye,Y.H.andTam,J.P.(Eds.)Peptides:Biology and Chemistry(Proceedings of the 1996 China Peptides Symposium),Kluwer Academic Publishers,The Netterlands, 1998, P.106.
2. (Wang. J.M., Hou, T.J. and Xu, X.J., Chemometrics and Intelligent Laboratory Systems, unpublished).
3. Kvasnicka, V. and Pospichal, J., J. Chem. Inf. Comput. Sci., 34 (1994) 1109.
4. David, R.W., Jounal of Computer-Aided Molecular Design, 11 (1997) 209.
5. Gehlhaar, D.K., Chem. Bio., 2 (1995) 317.

Aggregation of a calmodulin-binding peptide

Hu-Sheng Yan, Xiao-Hui Cheng, Ai-Guo Ni and Bing-Lin He

Institute of Polymer Chemistry, The State Key Laboratory of Functional Polymer Materials for Adsorption and Separation, Nankai University, Tianjin, 300071, China

Introduction

It is well known that protein folding is important in biological processes. In the folded protein the non-polar amino acid side chains are removed from water and packed together in the interior [1]. Understanding the molecular forces that determine protein folding still remains a problem in the physico-chemical aspects of protein research. It is appropriate to consider peptide aggregation as a model for protein folding, as the energetics involved in aggregation can throw light on the self-assembling ability of peptides, which plays a dominant role in protein folding [2, 3]. However, as peptides usually have great conformational flexibility, only a very limited number of examples have been reported to form ordered structures. In this paper, a peptide which forms highly ordered secondary structure is reported.

Results and discussion

According to a model for the calmodulin-binding domains of calmodulin-binding peptides [4, 5], a peptide with high affinity for calmodulin was designed and synthesized. The amino acid sequence of the peptide is: Asn-Ser-Arg-Lys-Lys-Leu-Leu-Leu-Leu- Leu-Leu-Gly-Ser-Gly-NH_2. The primary structure indicates that the peptide is basic and hydrophobic. The solubility of the peptide (as trifluoroacetic acid salt, which formed during the purification of the peptide by HPLC) was quite low in water and increased with increasing temperature. When the peptide in a warm aqueous solution at a concentration of c.a. 50 mg/ml or more was cooled to room temperature or lower, it became a transparent thermo-reversible gel. The result shows that the peptide strongly aggregated in aqueous solution. As the peptide is quite hydrophobic, addition of a low ratio of polar organic solvent to the solution should increase solubility and decrease the tendency to form a gel if the gel formation was caused by hydrophobic interaction. However, when ethanol or *n*-propanol (~10%) was added to the solution, the gel formed more quickly and became harder. The aggregation of the peptide in dilute aqueous solution, in which macroscopic gel formation was not occuring, was studied by dynamic light scattering. Fig. 1 shows the dynamic light scattering autocorrelation function of the peptide (as trifluoroacetic acid salt) in aqueous solution at a concentration of 13.0 mg/ml (pH ~3). The cumulants method [6] was used for the data analysis. The average diffusion coefficient, D_T, of the particles in the solution was calculated to be 1.04×10^{-8} cm^2/s. The corresponding average hydrodynamic radius was then calculated to be 331 nm using Stokes-Einstein relation ($R_h = kT/(6\pi\eta D_T)$, where k is the Boltzmann's constant and η is the viscosity of the solvent at the absolute temperature T), indicating that the peptide strongly aggregated to form polymer even in dilute solution. The polydispersity parameter was calculated to be 0.31, suggesting a polydisperse distribution of the particles in the solution.

Table 1 shows the effect of the peptide concentration and temperature on the average hydrodynamic radii of the particles in the solution. It can be seen that the radii of the aggregates decrease with increasing temperature and decreasing concentration of the peptide, respectively, in agreement with the effect of concentration and temperature on macroscopic gel formation. Particles were larger at 0.11 mg/ml than at 1.20 mg/ml, which may be caused by dust contributing more significantly to the average particle radius at the lower concentration.

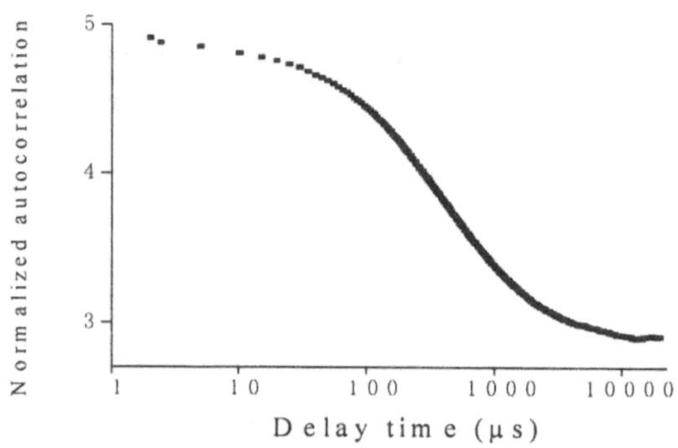

Figure 1. Autocorrelation function of the peptide at a concentration of 13.0 mg/ml at 25°C. (Laser light wavelength is 514.5 nm. Scattering angle is 90°.)

Many natural biopolymers with the ability to form a gel in aqueous solution, e.g. agarose and gelatin, have one essential feature in common, i.e. that the gelation is preceded by a transition from random coils to helical conformations [7]. We studied the conformation of the peptide in aqueous solution by infrared spectroscopy (IR). Fig. 2 shows the IR spectrum of the peptide in D_2O at a concentration of ~30 mg/ml. The strong absorption band at 1622 cm^{-1} and shoulder at ~1680 cm^{-1} can be attributed to the antiparallel β-pleated sheet structure, which is maintained by strong hydrogen bonds. The lower frequency of this absorption maximum confirms the existence of intermolecular interactions. The small absorption peak at ~1645 cm^{-1} shows that the content of the α-helical and disordered structures was quite low. The strong absorption at 1673 cm^{-1} was most probably caused by trifluoroacetic acid [8]. The above results show that the secondary structure of the peptide in aqueous solution was mainly antiparallel β-pleated sheets. The mechanism of the gel formation of the peptide is different from that of the natural biopolymers mentioned above.

Table 1. The average hydrodynamic radii (nm) of the particles determined by dynamic light scattering.

Concentration (mg/ml)	4°C	15°C	25°C	40°C	60°C
13.0	407	359	331	342	324
1.20	203	164	162	163	166
0.11	285	253	244	241	252

Usually, electrostatic, hydrogen-bonding and hydrophobic interactions are believed to contribute to peptide aggregation. As the peptide studied in this paper does not contain acidic groups, electrostatic interaction should not contribute to its aggregation. The association of two linear peptides to an isolated β-sheet element is energetically unfavourable, as the energy contribution of the stabilizing hydrogen bonds is not sufficient to overcome the loss of entropy due to the restriction of rotations of the peptide chains [9]. Therefore, hydrophobic interaction must have stabilized the β-pleated sheets. Generally speaking, the hydrophobic interaction has been most frequently ascribed as the determinate of the aggregation phenomenon. The free energy decrease due to the hydrophobic interaction is usually much greater than the decrease due to the hydrogen-bonding interaction in peptide aggregation or protein folding [10-12]. In addition, the peptide studied in this paper is quite hydrophobic. Thus, the favourable free energy of the aggregation may be caused largely by the hydrophobic interaction.

Addition of a low ratio of organic solvent to aqueous solutions of peptides or proteins should enhance the hydrogen-bonding interaction and significantly weaken the hydrophobic interaction. We can conclude that the hydrogen-bonding interaction plays a role in the gel formation since the addition of the organic solvent to the aqueous solution of the peptide was favourable for gelation. In other words, the free energy decrease due to the hydrophobic interaction may be much higher than that required by the gel formation, but the changes of the free energy within a certain range would not influence the characteristic of the aggregates.

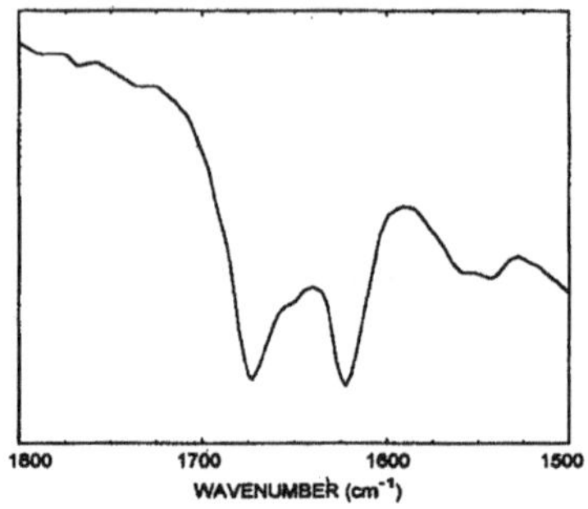

Fig. 2. IR spectrum of the peptide in D_2O solution.

Acknowledgments

This work was supported by National Natural Science Foundation of China.

References

1. Kauzmann, W., Adv. Protein Chem., 14 (1959) 1.
2. Privalov, P.L. and Gill, S.J., Pure Appl. Chem., 61 (1988) 1097.
3. Zhang, S., Holmes, T., Lockshin, C. and Rich, A., Proc. Natl. Acad. Sci. USA, 90 (1993) 3334.
4. Yan, H.S. and He, B.L., Chinese J. Biochem. Biophys., 24 (1992) 295.
5. Yan, H.S., Ni, A.G., Liu, L.P., Cheng, X.H. and He, B.L., Sci. China, Ser. B, 39(1996) 113.
6. Koppel, D.E., J. Chem. Phys., 57 (1972) 4814.
7. Viebke, C., Piculell, L. and Nilsson, S., Macromolecules, 27 (1994) 4160.
8. Li, S.C., Kim, P.K. and Deber, C.M., Biopolymers, 35 (1995) 667.
9. Williams, D.H., Aldrichim. Acta, 24 (1991) 71.
10. Mahaney, J.E. and Thomas, D.D., Biochemistry, 30 (1991) 7171.
11. Collawn, J.F. and Paterson, Y., Biopolymers, 29 (1990) 1289.
12. Kyte, J. and Doolittle, R.F., J. Mol. Biol., 157 (1982) 105.

Session C
Neuro/Endocrino/Bioactive peptide

Chairs: John Wade
Howard Florey Institute
University of Melbourne
Parkville, Australia

Meng-Shen Cai
School of Pharmaceutical Sciences
Beijing Medical University
Beijing, China

Neuropeptide AVP(4-8) playing an indirect neurotrophic role

Qing-Wu Yan, Xiao-Guang Zhen, Tong-Xi Wang and Yu-Cang Du
Shanghai Institute of Biochemistry, Chinese Academy of Sciences, Shanghai, 200031, China

Introduction

Arginine-vasopressin (AVP), one of the neurohypophyseal hormones, not only has peripheral effects on the regulation of body fluid volume and osmolality as well as maintenance of blood pressure but also has a central effect as it can enhance the memory behavior of the rat. In 1983, Burbach *et al* found that AVP(4-8), an endogenous metabolite of AVP, did not show any of the peripheral functions of AVP, but had a more potent central effect than its parent peptide [1]. Since the 1990's, our research group has done a series of studies to investigate the central functions of this peptide in rat brain as well as in a cultured cell line. AVP(4-8) may play a trophic role in the central nervous system by enhancing the expression of neurotrophic factors such as NGF and BDNF [2, 3]. Experiments from a cultured cell line demonstrated that AVP(4-8) could markedly stimulate the growth of C6 glioma cells in the early stage of cell culture [4]. It could also affect the apoptosis of C6 and compromise the increase of CDK activity. All these results implied that AVP(4-8) could be regarded as a neurotrophic factor in rat brain, but the mechanism of its trophic functions remains unclear.

Results and Discussion

There has been more and more evidence supporting the contention that neuropeptides could function as neurotrophic factors in the central nervous system. Neuropeptides have been demonstrated to affect a series of parameter in cultured CNS-derived cells, including cell division, neuronal survival, neurite sprouting and growth cone motility, as well as neuronal and glial phenotype. Some of these actions are direct, whereas others are mediated indirectly, such as when a neuropeptide stimulates an astrocyte to produce a neurotrophic factor needed for neuron survival.

Neuroglial cells play important roles in the central nervous system. One of these functions is providing neurotrophic factors for neurons to support their growth and development. Essentially every neuropeptide receptor has been found on glia, implying that neuropeptides can function as indirect trophic factors for neurons through regulating the expression of trophic factors.

The C6 glioma line was originally cloned from an N-nitrosomethyl-urea-induced rat glioma [5]. C6 has been used extensively as an *in vitro* glial model system due to its normal glia-like properties, including the secretion of some neurotrophic and laters neurite-promoting factors such as nerve growth factor and laminin [6]. We have reported that AVP(4-8) markedly stimulated cell growth in the early stage, and later, apoptosis of C6 in a concentration-dependent manner. Was its function a direct one or indirect one, i.e. did AVP(4-8) enhance C6 releasing some trophic factors, and then these factors fed back to C6 itself? To answer this question, we examined the effects of C6 conditioned medium of on the morphology of PC12 cells.

PC12 is a clonal cell line derived from rat pheochromocytoma. When cultured in the presence of neurotrophic factor, these cells extend long branching neurites and undergo several biochemical changes characteristic of sympathetic neurons [7]. Thus, it has been widely used to investigate the mechanisms underlying neuronal differentiation. In our experiments, C6 cells were cultured in serum-free medium. Following treatment with AVP(4-8) for five days, the conditioned media were collected and added to the serum-free medium in PC12 culture. We found AVP(4-8)-treated C6 conditioned medium (AVP(4-8)-C6-CM) could enhance the development of PC12 cells from a phenotype of pheochromocytoma to that of sympathetic neurons. The results showed that AVP(4-8)-C6-CM could induce PC12 development into the typical sympathetic neuron and markedly increase the number of neurites-bearing cells compared with untreated C6-CM. On the other hand, its parent peptide AVP and its antagonist ZDC(C)PR showed no effects. Furthermore, directly adding AVP(4-8) or C6-CM plus AVP(4-8) to PC12 culture showed no effect on the neurites growing out of these cells. Experiments also demonstrated that the AVP(4-8)-C6-CM had a neurite-promoting effect in a dilution-dependent manner. These results suggest that some unknown neurotrophic factors were secreted by AVP(4-8)-treated C6 cells to the medium which can function as enhancers for PC12 differentiation.

In summary, our experiments demonstrated directly that AVP(4-8) acted as a secretagogue in releasing a neuron development-promoting factor. The results also suggest a communications network that connects neurons not only directly but also indirectly via the non-neuronal cells.

References

1. Burbach, J.P.H., Kovacs, G.L. and de Wied, D., Science, 221 (1983) 1310.
2. Zhou, A.W., Guo, J., Wang, H.Y., Gu, B.X. and Du, Y.C., Peptides, 16 (1995) 1.
3. Zhou, A.W., Li, W.X., Guo, J. and Du, Y.C., Peptides, 18 (1997) 1179.
4. He, M., Miao, H.H., Chen, X. F. and Du, Y. C., Chin. J. Cell. Biol., 17 (1995) 176.
5. Benda, P., Lightbody, J., Sato, G., Levine, L. and Sweet, W., Science, 161 (1968) 370.
6. Westermann, R., Mollenhausen, J., Johanssen, J. and Unsicker, K., Int. J. Dev. Neurosci., 7 (1988) 219.
7. Greene, L.A. and Tischler, A.S., Proc. Natl. Acad. Sci. USA, 73 (1976) 2424.

Potent bradykinin antagonists having medical potential

John M. Stewart[a, b], Lajos Gera[a], Eunice J. York[a] and Daniel C. Chan[b]

[a]*Department of Biochemistry and* [b]*Cancer Center, University of Colorado School of Medicine, Denver, CO 80262, USA*

Introduction

The nonapeptide bradykinin (Arg-Pro-Pro-Gly-Phe-Ser-Pro-Phe-Arg; BK) and the decapeptide kallidin (Lys-BK) are involved in the normal regulation of every major physiological system [1]. Examples are regulation of blood pressure and organ perfusion, stimulation of ion transport (including sodium excretion), contraction or relaxation of smooth muscles in the gastrointestinal and respiratory systems, and stimulation of growth, particularly wound repair. Actions of BK in different tissues can involve every known second messenger system; prominent among these are phospholipase A_2 (PLA_2) and phospholipase C (PLC). In vascular tissue, endothelial BK receptors are linked to nitric oxide synthase (NOS); the NO diffuses into the smooth muscle to stimulate guanylate cyclase and produce cGMP for muscle relaxation.

Actions of BK in pathophysiology are equally diverse and impressive. BK is the initiator – or an early mediator – of nearly all inflammation [2]. BK can produce all the cardinal signs of inflammation – calor (fever), rubor (redness due to vasodilatation), dolor (pain) and tumor (edema); in serious inflammation it can lead to loss of function and multiple organ failure. Many bacteria release enzymes that directly or indirectly cause BK production[3]. The subsequent BK-evoked increased vascular permeability facilitates diffusion of the bacteria to spread the infection. Similarly, the inflammatory reaction produced by solid tumors facilitates tumor enlargement and metastasis. BK has potent growth factor activity; some tumors, notably small cell lung carcinoma (SCLC), produce BK and use it as an autocrine growth stimulant [4]. BK is the most potent known nociceptive agent; the pain it evokes is an important part of inflammation. Stimulation of PLA_2 releases arachidonic acid for prostaglandin production, further exacerbating pain. These actions of BK show it to be a major agent in many serious conditions, such as shock, asthma, adult respiratory distress syndrome (ARDS), inflammatory joint disease (arthritis), inflammatory bowel disease and consequences of major trauma.

Production of BK can be initiated in many ways. The "contact activation system" is perhaps the best known [1]. In contact activation, BK is produced whenever the blood clots due to contact with a negatively charged surface, such as collagen or basement membrane. Active plasma kallikrein, the enzyme that produces BK, is produced by activated Factor XII of the blood clotting cascade. Pre-kallikrein can also be activated by low pH, such as occurs in ischemia when circulation is compromised. Poor neurological outcome of carotid entarterectomy has been linked to BK production in the cerebral circulation [5]. Many glands – notably pancreas, kidney and salivary glands – contain the second kinin-forming enzyme precursor for glandular (or tissue) kallikrein. This enzyme produces kallidin upon activation and release and appears to be an important participant in acute pancreatitis.

Inactivation of BK normally occurs very rapidly in the circulation. Angiotensin I converting enzyme (ACE) destroys nearly 100% of the BK passing through the pulmonary circulation by removal of the C-terminal dipeptide. Any BK that escapes cleavage by ACE is hydrolyzed by plasma carboxypeptidase N (CPN), which removes the C-terminal arginine residue. Aminopeptidase P (APP) removes the N-terminal arginine, and neutral endopeptidase (NEP) cleaves BK at 5-phenylalanine. These enzymes generally keep BK concentration under control, but during massive overproduction of BK or loss of a cleaving enzyme, excess BK can cause major consequences, such as shock and serious pain.

Two receptors, B1 and B2, mediate the actions of kinins. The B2 receptor is constitutively expressed in most tissues, and requires the full BK chain for agonist activity. None of the products of the kininase enzymes can activate B2 receptors. In contrast, B1 receptors are normally not expressed, but are evoked rapidly in inflammation. The normal ligands for B1 receptors are the products of CPN action: BK(1-8) and kallidin(1-9). In sepsis, ACE is lost from the lungs, kinins are processed by CPN, and activation of the concomitantly expressed B1 receptors by these shortened peptides leads to the hypotensive crisis of shock. Activation of B1 receptors can also help counteract shock since these receptors are strongly linked to release of pressor catecholamines.

Results and Discussion

Involvement of kinins in such varied and serious pathologies suggests that effective kinin antagonists should be useful drugs. Shortly after the discovery of B1 receptors in 1977 it was found that replacement of the C-terminal Phe in BK(1-8) or kallidin(1-9) by leucine gave an antagonist for these receptors [6], but not for B2 receptors. The major impact on the kinin field, however, was discovery in 1984 of B2 antagonists in which the Pro^7-residue was replaced by D-phenylalanine [7]. This D-amino acid residue blocked inactivation by ACE, but cleavage by CPN converted these B2 antagonists into B1 antagonists [8]. Nevertheless, these “first generation” antagonists, such as **NPC-349** (DArg-Arg-Pro-Hyp-Gly-Thi- Ser-D-Phe-Thi-Arg; Hyp=4-hydroxyproline, Thi= β-2-thienylalanine), were very important tools for definition of the scope and importance of kinin physiology and pathology.

The “second generation” of BK antagonists was introduced by the Hoechst **HOE-140** (Icatibant; DArg-Arg-Pro-Hyp-Gly-Thi-Ser-DTic-Oic-Arg; Tic=tetrahydro-isoquinoline -3-carboxylic acid; Oic=octahydroindole-2-carboxylic acid) [9]. The most important feature of **HOE-140** is the Oic^8 residue, which prevents cleavage by CPN and gives an antagonist specific for B2 receptors that has a very long lifetime *in vivo*. Also in the "second generation" is Bradycor, introduced by Cortech. It is a dimer of a Cys6 first generation antagonist (DArg-Arg-Pro-Hyp-Gly-Phe-Cys-DPhe-Leu-Arg) in which the crosslink connecting the two Cys residues is *bis*-maleimidohexane (BSH) [10]. Both Icatibant and Bradycor have been used in limited clinical trials for conditions such as asthma, sepsis and head trauma.

The "third generation" of BK antagonists came with our introduction of peptides containing α-(2-indanyl)glycine (Igl) [11]. One of the best of these is **B-9430** (DArg-Arg-Pro-Hyp-Gly-Igl-Ser-DIgl-Oic-Arg). The Igl residue at position 5 prevents cleavage by NEP and gives a peptide that is stable indefinitely in the circulation and in tissue homogenates. It is orally available in rats, and a single subcutaneous injection can block BK action for two days [12]. The properties of this peptide suggest it should be clinically useful for conditions involving both acute and chronic inflammation.

A further significant increase in potency of BK antagonists (fourth generation?) came with introduction of pentafluorophenylanine (F5f) [13]. Compound **B-10056** (DArg-Arg-Pro-Hyp-Gly-Igl-Ser-DF5f-Igl-Arg) is an order of magnitude more potent than **B-9430** in the rat uterus assay. Extensive biological studies have not yet been done with these new peptides.

Although there is significant interest in BK antagonists in the pharmaceutical industry, most research is now directed toward non-peptide BK antagonists. The first important compound is the Fujisawa antagonist **FR173657**, which is potent, specific, and is orally available [14]. Limited information is also available on a Pfizer antagonist [15], and compounds from other companies can be anticipated.

An important target for BK antagonists is small cell carcinoma of the lung [16]. For years we have tested all our BK antagonists against small cell and non-small cell cancers in *in vitro* cell culture. The BK receptors on SCLC are B2 and are linked to elevation of intracellular free calcium concentrations. All our potent antagonists, such as **B-9430**, show marked inhibition of this BK-evoked calcium elevation, but none of these peptides inhibited cell growth. In an attempt to develop cytotoxic activity, we crosslinked potent antagonists at the N-terminus with various linkers. One of the best of these dimers is **B-9870**, in which two molecules of **B-9430** are linked by a seven-carbon diimide linker (suberimide; SUIM). This dimer shows an *in vitro* IC_{50} of 0.15 μM against the SCLC strain SHP77 [17]. Other laboratories have reported cytotoxicity of antagonists of bombesin and substance P (SP) against these cells, but in our comparison, SP antagonists had much lower potency which was not improved by dimerization. The Cortech dimer Bradycor also had very low potency in this assay (IC_{50}=50 μM). A SUIM dimer of **HOE-140** had an IC_{50} of 35 μM. **B-9870** showed impressive inhibition of tumor growth *in vivo* in nude mice bearing implants of the same tumor strain (SHP77) when administered intraperitoneally at a daily dose of 5 mg/kg.

Conclusion

Bradykinin antagonist peptides have been developed during the past fourteen years into extremely potent and stable molecules that offer great potential for development into drugs for serious inflammation and cancers. The great need now is for financial support for these human studies.

Acknowledgments

We thank Vikas Dhawan, Paul Bury, Marcos Ortega and Frances Shepperdson for technical assistance, and the US NIH for grants HL-26284 and CA-78154.

References

1. Bhoola, K., Figueroa, C.D. and Worthy, K., Pharmacol. Rev., 44 (1992) 1.
2. Stewart, J.M., Agents Actions, 42 (1993) 145.
3. Maeda, H., Akaike, J.W., Noguchi, Y. and Sakata, Y., Immunopharmacology, 33 (1996) 222.
4. Bunn, P.A., Chan, D., Stewart, J., Gera, L., Tolley, R. Jewett, P., Tagawa, M., Alford, C., Mochizuki, T. and Yanaihara, N., Cancer Res., 54 (1994) 3602.
5. Makevnina, L.G., Lomova, I.P., Zubkov, Y.N. and Semenyutin, V.B., Brazil. J. Med. Biol. Res., 27 (1994) 955.
6. Regoli, D., Barabe, J. and Park, W.K., Can. J. Physiol. Pharmacol., 55 (1977) 855.
7. Vavrek, R.J. and Stewart, J.M., Peptides 6 (1985) 161.
8. Regoli, D., Drapeau, G., Rovero, P., Dion, S. and D'Orleans-Juste, P., Eur. J. Pharmacol., 123 (1986) 61.
9. Hock, F.J., Wirth, K., Albus, U., Linz, W., Gerhards, H.J., Wiemer, G., Henke, St., Breipohl, G., König, W. and Schölkens, B.A., Br. J. Pharmacol., 102 (1991) 769.
10. Cheronis, J.C., Whalley, E.T., Nguyen, K.T., Eubanks, S.R., Allen, L.G., Duggan, M.J., Loy, S.D., Bonham, K.A. and Blodgett, J.K., J. Med. Chem., 35 (1992) 1563.
11. Stewart, J.M., Gera, L., Hanson, W., Zuzack, J.S., Burkard, M., McCllough, R. and Whalley, E.T., Immunopharmacology, 33 (1996) 51.
12. Whalley, E.T., Hanson, W.L., Stewart, J.M. and Gera, L., Can. J. Physiol. Pharmacol., 75 (1997) 629.
13. (Gera, L. and Stewart, J.M., In Tam, J.P. (Eds.) Proc. 15th Amer. Peptide Symp., 1998, unpublished.)
14. Inamura, N., Asano, M., Kayakiri, H., Hatori, C., Oku, T. and Nakahara, K., Can. J. Physiol. Pharmacol., 75 (1997) 622.
15. Ikeda, T., Chem. Abstr., 127 (1997) 248125.
16. Woll, P.J. and Rozengurt, E., Br. J. Cancer, 57 (1988) 579.
17. Chan, D., Gera, L., Helfrich, B., Helm, K., Stewart, J., Whalley, E. and Bunn, P., Immunopharmacology, 33 (1996) 201.

Understanding the chemistry and biology of rodent relaxin: Role of the C-peptide

John D Wade[a], Nicola F Dawson[a], Mary Macris[a], Marc Mathieu[a], Yean-Yeow Tan[b], Roger J Summers[b] and Geoffrey W Tregear[a]

[a]*Howard Florey Institute, University of Melbourne, Parkville, Victoria 3052, Australia and* [b]*Department of Pharmacology, Monash University, Clayton, Victoria 3168, Australia*

Introduction

Relaxin, a two-chain, three disulfide-bonded member of the insulin superfamily, is produced principally by the corpus luteum of the ovary during pregnancy. It has long been considered that its primary role is that of preparing the birth tract for the delivery of the young [1]. However, the recent demonstration of binding sites for relaxin in the heart and brain of both male and female rats points to a wider physiological role [2, 3]. Studies with synthetic relaxin have shown the peptide possesses powerful chronotropic and inotropic activity [4] and to directly increase neuronal activity in regions of the lamina terminalis and hypothalamus associated with fluid and electrolyte balance [5].

Like insulin, relaxin is made on the ribosome as a preprohormone which undergoes subsequent proteolytic processing to yield the native two-chain peptide. Curiously, the connecting C-peptide is approximately 103 residues in length, compared to the corresponding 25 residue peptide in proinsulin [6]. Studies with insulin have shown that C-peptide mimics as short as one residue are sufficient to effect efficient chain folding and correct disulfide bond pairing [7]. Furthermore, recombinant DNA-derived human Gene 2 relaxin may be readily and efficiently produced using an "ini" C-peptide of 13 residues. This C-peptide is then excised by treatment with specific endopeptidases [8]. Thus, the large length of the native relaxin C-peptide suggests that it may perhaps have an additional function or serve as a precursor to biologically active peptides. Support for this concept has recently arisen following the surprising finding that, in pharmacological doses, human proinsulin C-peptide has an insulin-like action [9, 10].

We therefore undertook the chemical synthesis of two fragments of rat relaxin C-peptide and assayed these for both relaxin-like or relaxin-potentiating activity.

Results and Discussion

The primary structure of rat prorelaxin is shown in Fig 1. The 240 residue A-chain is linked to the 35 residue B-chain by a C-peptide that is 105 residues long. The pair of basic residues shown in bold italics within the C-peptide sequence is known to be invariant in all known relaxin C-peptides. It is thought that proteolytic processing of prorelaxin would also result in cleavage at this site to yield two fragments, a 73 residue N-terminal peptide (1-73) and a 26 residue C-terminal segment (76-101).

B-chain

R-V-S-E-E-W-M-D-Q-V-I-I-Q-V-C-G-R-G-Y-A-R-A-W-I-E-V-
C-G-A-S-V-G-R-L-A-L-S-Q-E-E-P-A-P-L-A-R-Q-A-T-A-E-V-
V-P-S-F-I-N-K-D-A-E-P-F-D-M-T-L-K-C-L-P-N-L-S-E-E-R-
K-A-A-L-S-E-G-R-A-P-F-P-E-L-Q-Q-H-A-P-A-L-S-D-S-V-V-
S-L-E-G-F-**K-K**-T-F-H-N-Q-L-G-E-A-E-D-G-G-P-P-E-L-K-Y-
L-G-S-D-A-Q-S-R-K-K-R-**Q-S-G-A-L-L-S-E-Q-C-C-H-I-G-C-**
T-R-R-S-I-A-K-L-C **A-chain**

Fig. 1. Primary structure of rat prorelaxin. Italics show the sequence of the connecting C-peptide.

The solid phase synthesis of the two fragments of rat relaxin C-peptide was achieved using the continuous flow Fmoc methodology as previously described [11, 12]. Fmoc-amino acid O-pentafluorophenyl esters were used with the exception of arginine, which was coupled as its HBTU-activated species. N^{α}-Fmoc deprotection was with 20% piperidine in DMF. Amino acid side chain protection was afforded by the following:Arg, Pmc; Asn and Gln, Trt; Asp and Glu, Bu^t; Cys, Trt; His, Trt; Lys, Boc; Ser and Thr, Bu^t. For the assembly of the shorter peptide, use of a single N-(2-hydroxy-4-methoxybenzyl) (Hmb) amide bond protecting group [13] prevented possible base-mediated aspartimide formation at the Asp-Gly at positions 87-88 [14]. For both peptides, no repeat amino acid couplings were required and at the end of assembly, cleavage from the solid supports and side chain deprotection was achieved by extended treatment of the two peptide-resins with TFA in the presence of scavengers.

Both peptides were subjected to purification by conventional preparative RP-HPLC on Vydac C18 supports using TFA-based buffers. Overall yields of the two peptides were approximately 7 and 30%, respectively. The resulting products were subjected to comprehensive chemical characterization including analytical RP-HPLC (fig. 2), and capillary zone electrophoresis. Matrix assisted laser desorption time-of-flight mass spectrometry (MALDITOF MS) analysis of the two peptides gave values close to these calculated [C-peptide (1-73): found MH+ 7,904.0; calc. MH+ 7907.8; C-peptide (76-101): found MH+ 2,761.8; calc. MH+ 2,761.9].

The two synthetic peptides were initially assayed for characteristic relaxin activity in the rat isolated atrial assay. This provides an extremely sensitive and qualitative measure of the chronotropic and inotropic activity of relaxin peptides. Other members of the insulin superfamily including insulin and insulin-like growth factor I are devoid of activity in this assay even in pharmacological doses [15]. Remarkably, porcine prorelaxin was shown to be as active as native porcine relaxin in this assay suggesting that the presence of the C-peptide may not affect the binding of the peptide to the receptor [15].

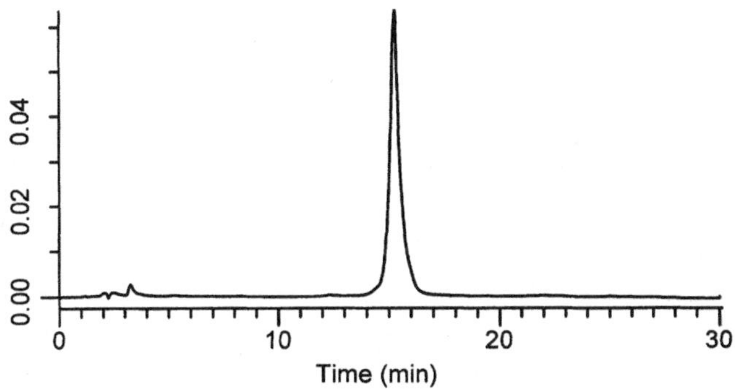

Fig. 2. RP-HPLC of purified synthetic rat relaxin C-peptide (1-73). Column: Vydac C18. (analytical). Buffer A: 0.1% aq. TFA, Buffer B: 0.1% TFA/CH_3CN. Gradient: 20-40%B, 30 min. Wavelength: 214 nm. Flow rate: 1.5 ml/min.

As shown in Fig. 3 (black bars), human Gene 2 relaxin (A24/B29) caused a powerful increase in both the rate and strength of atrial contraction when administered at a concentration of either 1 or 10 nM. In contrast, the synthetic rat C-peptide (1-73), at a concentration of 1 μM, was devoid of chronotropic and inotropic activity even after 15-20 min of incubation. When either 1 or 10 nM human Gene 2 relaxin (A24/B29) was added to the organ bath, its chronotropic and inotropic responses were neither significantly augmented nor inhibited (white bars). A similar result was obtained with the synthetic rat C-peptide (76-101) (data not shown).

The function of the C-peptide (if any) thus far remains unknown. The full-length 101-residue peptide is presently being assembled. Together with the two synthetic C-peptide fragments, it will be screened for novel biological activities. The recent detection by radioimmunoassay of high levels of relaxin C-peptide in seminal plasma [16] suggests another potential role for the peptide. Specific antibodies currently being produced against the peptides may assist in more clearly defining its role other than in folding and oxidation of the relaxin chains.

Acknowledgments

The work described herein was supported by an Institute Block grant from the National Health and Medical Research Council of Australia.

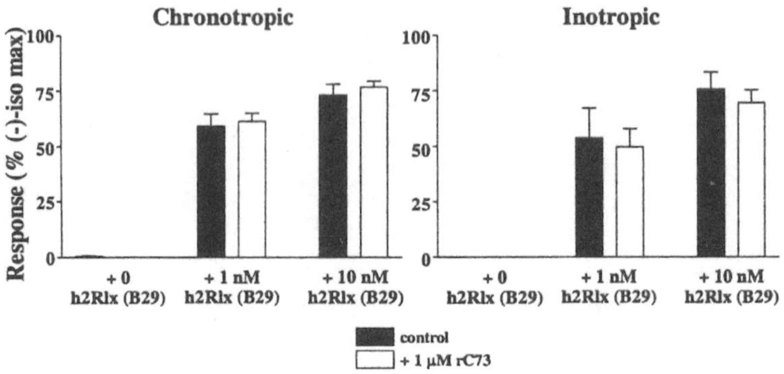

Fig. 3. Isolated rat atrial assay of synthetic rat relaxin C-peptide (1-73) [rC73]. h2Rlx (B29) = human Gene 2 relaxin (A24/B29). N=6. Control = 0.1% aq.TFA. Response=% of maximal response produced by (-)-isoprenaline.

References

1. Bryant-Greenwood, G.D. and Schwabe, C., Endocrine Rev., 15 (1994) 5.
2. Osheroff, P.L., Cronin, M.J. and Lofgren, J.A., Proc. Natl. Acad. Sci. USA, 89 (1992) 2384.
3. Osheroff, P.L. and Phillips, H.S., Proc. Natl. Acad. Sci. USA, 88 (1991) 6413.
4. Kakouris, H., Eddie, L.W. and Summers, R.J., Lancet, 339 (1992) 1076.
5. McKinley, M.J., Burns, P., Colvill, L.M., Oldfield, B.J., Wade, J.D., Weisinger, R.S. and Tregear, G.W., J. Neuroendocrinol., 9 (1997) 431.
6. Sherwood, O.D., In Knobil, E. and Neill, J.D. The Physiology of Reproduction. 2nd edition. Raven Press, New York, USA, 1994, p. 861.
7. Busse, W.D. and Carpenter, F.H., Biochem., 15 (1976) 1649.
8. Vandlen, R., Winslow, J., Moffat, B. and Rinderknecht, E. In MacLennan, A.H., Tregear, G.W. and Bryant-Greenwood, G.D. (Eds.) Progress in Relaxin Research (Proceedings of the 2nd International Congress on the Hormone Relaxin), Global Publication Services, Singapore, 1995, p.59.
9. Ido, Y., Vindigni, A., Chang, K., Stramm, L., Chance, R., Heath, W.F., DiMarchi, R.D., Di Cera, E. and Williamson, J.R., Science, 277 (1997) 563.
10. Forst, T., Kunt, T., Pohlman, T., Goitom. K., Englembach, M., Beyer, J. and Pfzner, A., J. Clin. Invest., 101 (1998) 2036.
11. Atherton, E. and Sheppard, R.C., Solid Phase Peptide Synthesis. IRL Press at Oxford University Press, Oxford, United Kingdom, 1989.

12. Wade, J.D., Lin, F., Talbo, G., Otvos, L., Tan, Y.Y. and Summers, R.J., Biomed. Peptides Proteins Nuc. Acids, 2 (1997) 89.
13. Johnson, T., Quibell, M. and Sheppard, R.C., J. Peptide Sci., 1 (1995) 11.
14. Quibell, M., Owen, D., Packman, L.C. and Johnson, T., J. Chem. Soc., Chem. Commun., (1994) 2343.
15. Tan, Y.Y., Wade, J.D., Tregear, G.W. and Summers, R.J., Brit. J. Pharmacol., 123 (1998) 762.
16. Borthwick, G.M., Borthwick, A.C., Grant, P. and MacLennan, A.H., In MacLennan, A.H., Tregear, G.W. and Bryant-Greenwood, G.D. (Eds.) Progress in Relaxin Research (Proceedings of the 2nd International Congress on the Hormone Relaxin), Global Publication Services, Singapore, 1995, p.251.

Purification and characterization of a new conotoxin from the venom of *Conus betulinus*

Chong-Xu Fan, Ming-Nai Zhong, Hui Jiang, Shang-Yi Liu, Da-Yu Li, Min Liu, Shou-Lan Zhang, Fu-Sheng Lin and Ji-Sheng Chen

Research Institute of Pharmaceutical Chemistry, Beijing, 102205, China

Introduction

Marine snails belonging to the genus *Conus* are venomous predators that have developed a unique biochemical strategy for envenoming their prey. Their venom contains a number of small peptides ranging in size from ~13 to 30 amino acid residues. The structures and mode of action of many peptides from *Conus* venom have been elucidated [1, 2]. Among these, are the α-conotoxins that inhibit nicotinic acetylcholine receptors, the δ- and μ-conotoxins that act on voltage-sensitive sodium channels, and ω-conotoxins which selectively inhibit the neuronal subtypes of voltage-sensitive calcium channels. Up to now, most work on *Conus* venom concerned the fish-hunting species (*C. geographus*, *C. magus*, *C. striatus* etc.) that have seriously poisoned humans. However, the genus *Conus* is composed of approximately 500 species, and the fish-hunting species are in the minority (about 10%). The largest numbers of species are believed to be worm-hunting. Although all *Conus* species are venomous, the worm-hunting species have been seldom studied. *Conus betulinus*, a worm-hunting species, is one of the most abundant species in the south sea of China. In our previous work [3], we found that the crude venom of *C. betulinus* induces a series of symptoms such as aggressiveness, round movement, stiff tail, paralysis, convulsion and even death when injected intracisternally into mice. In this report we describe the purification and characterization of a peptide toxin from *C. betulinus* venom.

Results

Specimens of *C. betulinus* were collected from the coast of Sanya, Hainan Province, China. The venom ducts were dissected and extracted twice with 1.1% HOAc aqueous solution. All extracts were pooled, lyophilized and stored in the freezer.

The crude venom of *C. betulinus* was re-extracted with 1.1% HOAc. After centrifugation, the solution was subjected to gel filtration on a Sephadex G-25 column eluted with 1.1% HOAc. Five main fractions with different biological activities were obtained as shown in Figure 1A. The marked fraction causes paralysis of the hind legs of mice. It was lyophilized and separated by reverse phase HPLC on a semi-preparative C18 column eluted with a gradient of acetonitrile in 0.1% TFA (fig. 1B). The arrow labeled peak indicates the fraction containing peptide toxins. After further purification under the same condition, the pure peptide toxin BeIIIA obtained. Its purity was monitored by HPCE. Bioassay indicated that conotoxin BeIIIA induced stiff tail, circling movement and paralysis of hind legs when injected i.c. into mice at the dose of 0.5 mg/kg.

The amino acid composition of the peptide toxin determined by acidic hydrolysis indicated the peptide has 13 amino acid residues with calculated molecular weight of 1406 (Table 1). However, low-resolution fast atom bombardment mass spectrometry of native peptide showed the m/z of MH^+ was t 1592. The difference between the actual and calculated value was 185, which suggested the presence of another amino acid residue, tryptophan (186.2). Additionally, amino acid analysis by basic hydrolysis confirmed the presence of Trp (data not shown). The conotoxin BeIIIA thus contains 14 amino acid residues. The sequence of the peptide was determined by standard Edman method while cysteines were derived with acrylamide. The primary structure of BeIIIA was Cys-Cys-Lys-Gln-Ser-Cys-Thr-Thr- Cys-Met-Pro-Cys-Cys-Trp. The MALDI-TOF mass spectrum showed that the monoisotopic mass of the native peptide is thus 1589.3 Da, in good accordance with predicions from the sequence, with all cysteines in disulfide bonds and a free acid C-terminus (calculated 1589.5 Da). The complete sequence of BeIIIA is shown in Table 2 and is compared with peptides from another worm-hunting species, *C. quercinus* [3].

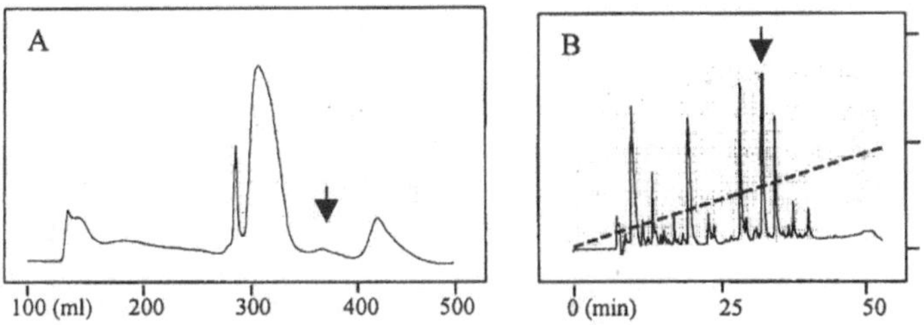

Fig. 1 (A) Chromatography of the crude venom on a Sephadex G-25 column (2.6x100 cm) equilibrated and eluted with 1.1% HOAc at a flow rate of 0.35ml/min, detected at 280 nm. The marked fraction contained the peptides causing mice paralysis. (B) Separation of the marked fraction by HPLC on a Phenomenex C18 semi-preparative column (1x25 cm, micron) eluted at a flow rate of 2 ml/min, detected at 214 nm. The gradient of acetonitrile in 0.1% trifluoroacetic acid is indicated by the dashed line.

Table 1. The amino acid composition of the conotoxin BeIIIA.

Amino acid	Thr	Ser	Glu	Pro	Cys	Met	Lys
Moles	2.0	1.0	1.2	1.2	5.6	1.0	1.0

Table 2. Comparison BeIIIA to QcIIIA and QcIIIB.

BeIIIA	**Cys-Cys**-Lys-Gln-Ser-**Cys**-Thr-Thr-**Cys**-Met-Pro-**Cys-Cys**-Trp (this work)
QcIIIA	**Cys-Cys**-Ser-Gln-Asp-**Cys**-Leu-Val-**Cys**-Ile-Hyp-**Cys-Cys**-Pro-Asn-NH2 (3)
QcIIIB	**Cys-Cys**-Ser-Arg-His-**Cys**-Trp-Val-**Cys**-Ile-Hyp-**Cys-Cys**-Pro-Asn (3)

As shown in Table 2, conotoxin BeIIIA is generally similar to conotoxin QcIIIA and QcIIIB in cysteine arrangement, containing three disulfide bonds. However, posttranslational modifications of amino acid are not seen in conotoxin BeIIIA. Biologically activities and action targets of conotoxin BeIIIA are being studied.

References

1. Olivera, B. M., Rivier, J., Clark, C., Ramilo, C. A., Corpuz, G. P., Abogadie, F. C., Mena, E. E., Woodward, S. R., Hillyard, D. R. and Cruz, L. J., Science, 248 (1990) 257.
2. Myers, R. A., Cruz, L. J., Rivier, J. E. and Olivera, B. M., Chem. Rev., 93 (1993) 1923.
3. Wei, K.H., Zhong, M.N., Chen, J.S. and Yu, L., In Xu, X.J., Ye, X.H. and Tam, J.P. (Eds.) Peptide: Biology and Chemistry (Proceedings of 1996 Chinese Peptide Symposium), Kluwer Academic Publishers, Dordrecht, 1996, p.169.
4. Abogadie, F. C., Ramilo, C. A., Corpuz, G. P. and Cruz, L. J., J. Trans. Natl. Acad. Sci. Tech. Philippines, 12 (1990) 219.

Synthesis and biological activity of human calcitonin analogs

Chao Yu, Da-Fu Cui, Guo-Ming Zhou, Bo-Liang Li and Xiang-Fu Wu

Shanghai Institute of Biochemistry, Academic Sinica, Shanghai, 200031, China

Introduction

Calcitonin is a 32 amino acid peptide hormone involved in the regulation of calcium metabolism. Calcitonins are found in a variety of vertebrate species including mammals, birds and fishes with considerable variability in amino acid sequences. Calcitonins of certain non-human species such as salmon calcitonin (sCT) and eel calcitonin (eCT) with 50% homology to human calcitonin (hCT) appear to be more potent in humans than hCT. In spite of their higher potency, sCT is not satisfactory for clinical use, because of its immunogenic property *in vivo* [1]. We are attempting to synthesize some hCT analogs which possess activities higher than the native one. Our modification is based on the studies of Basava C. [2]. It is known that the biological potency of CT is related to its comformational rigidity and flexibility [3] and the N-terminal of sCT is likely to form a more stable α-helix structure than hCT [4]. Our modification therefore focuses on the N-terminal of hCT. Some aromatic amino acids are substituted by hydrophobic amino acids to increase the conformational stability of hCT analogs. We have evaluated the potency of our synthetic $[V^0, L^{8, 12}]$hCT (hCT-SII) in osteoporosis therapy.

Results and Discussion

Solid phase synthesis of hCT-SII and its purification

ABI 430A peptide synthesizer and 0.2 mmole Boc-Pro-MBHA Resin were used. The N-terminal amino group was protected by Boc group. The carboxyl group of the protected amino acid was activated by HOBt through coupling with DCC. The peptide was cleaved from the resin by HF (5% p-cresol) at 0 °C for 90 min. and isolated by extraction with 80% acetic acid. The obtained linear peptide was dissolved in distilled water (1mg per 5ml) and subjected to air oxidation at pH7.5 for 30 hours. Through Sephadex G-10 desalting, and further purification by semi-preparative RP-HPLC, the purity of this peptide was > 95% as monitored by analytical HPLC and capillary electrophoresis (fig. 1). Its mass spectrum and N-terminal sequencing were in accordence with the theoretical values.

Biological assay in hypocalcemic activity of synthetic hCT-SII

The hypocalcemic activity of the synthetic hCT-SII was measured as follows: Male rats

(Wistar) weighing approximately 100 grams were fasted overnight. Peptide hCT-SII and sCT (Sigma, served as the standard) were dissolved in pH 4.0 acetate buffer and the potency was estimated by intraperitoneal method. The doses of each peptide corresponding to 30 ng and 9 ng at the higher level, and 10ng and 3ng at the lower level, respectively, were given in an appropriate volume. Two doses each of two different calcitonins were given to experimental groups of six rats. Six control rats were injected with acetate buffer. After 60 min the animals were anesthetized and the blood samples were taken from oculi. The calcium concentration in the serum was analyzed by the method of Xue JZ [5]. Calcium-lowering potency of the synthetic hCT-SII can be calculated by parallel line assay 2.2 (random) method and the potency of hCT-SII was around 2000 IU/mg, ten fold more than that of hCT (200 IU/mg). hCT-SII remained potent for about four hours after administration, while hCT remained potent about two hours, as reported by Habener [6]. In our studies using an RIA kit, the binding ability of hCT-SII to the antibody of hCT is even stronger than that hCT. These results indicated that the conformation of hCT-SII might have been changed, and its increased potency might depend on its greater resistance to metabolic degradation in vivo or its more favorable binding to the hormone receptor. Study of the structure of hCT-SII in solution is well under way.

Biological activity in rat osteoporosis models
Thirty male SD rats about 300g in weight and 12 weeks old were fed a low calcium diet (Ca: 0.1%, P: 0.6%) for six months. The rats were then fed a normal diet (Ca: 1.1%, P: 1.1%) and allocated randomly into 3 groups. One group with normal saline was used as the control and the other groups were treated, respectively, with hCT-SII or sCT in doses of 500ng and 200ng. The experimental animals were treated by subcutaneous injection of the peptides once every two days for three months. The parameters measured after treatment by scanning electronic microscopy (SEM) were bone density (fig. 2) and shape of bone trabecula. The results showed that the bone density in the two CT groups was significantly increased ($P<0.01$) compared with the control group. SEM showed that the bone trabecula of the treated groups was thick and continuous and that of the control group thin and discontinuous. This meant hCT-SII could enhance new bone formation by stimulation of osteoblasts while inhibiting bone resorption by altering osteoclastic and osteocytic activity. Thus, hCT-SII may play a promising role in osteoporosis therapy.

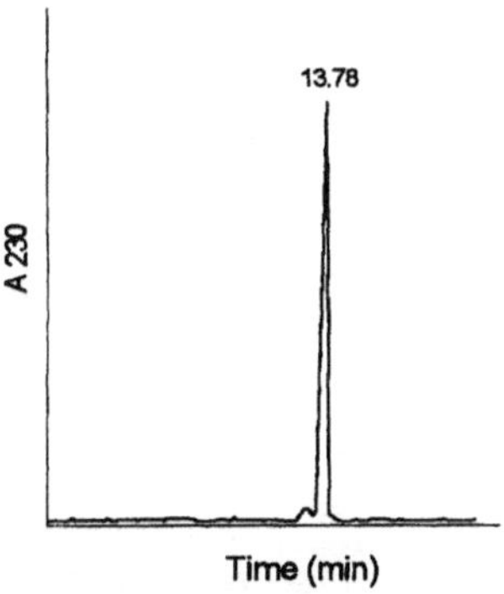

Fig. 1. Identifcation of hCT-SII with CE.

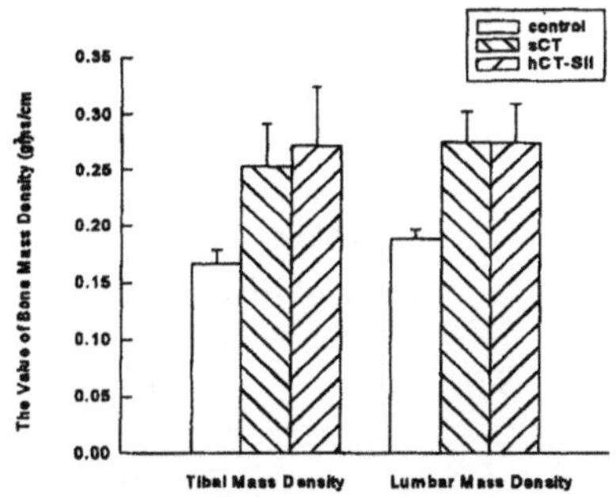

Fig. 2. Bone mass density of osteoprosis rats with therapy.

References

1. Muff, R., Dambacher, M.A. and Perrenoud, A., Am. J. Med., 89 (1990) 181.
2. Basava, C. and Hostetler, K.Y., In Smith, J.A. and River, J.E. (Eds.) Peptides: Chemistry and Biology (Proceedings of the 12th American Peptide Symposium), ESCOM, Leiden, 1992, p.20-22
3. Epand, R.M., Comprehensive Medicinal Chemistry, 3 (1990) 1023.
4. Arvinte, T., J. Biochem., 268 (1993) 6408.
5. Xue J.Z., Cao X.J. and Sun Z.Y., Chinese Journal of Medical Laboratory Technology, 7 (1984) 147.
6. Habener, Nature, 232 (1971) 91.

Enhanced expression of CNDF mRNA in rat brain by administration of AVP(4-8)

Wen-Xue Li, Ben-Xian Gu and Yu-Cang Du

Shanghai Institute of Biochemistry, Chinese Academy of Sciences, Shanghai, 200031, China

Introduction

AVP(4-8), also known as ZNC(C)PR, is a metabolite of arginine-vasopressin (AVP). It was found to be much more potent than AVP in facilitating the acquisition and maintenance of learning and memory in rats. Our previous work reported that ZNC(C)PR not only enhanced behavioral responses in rats, but also induced long-term potentiation (LTP). Furthermore, the gene expression of c-fos, NGF and BDNF was increased by administration of ZNC(C)PR [1]. NGF and BDNF are widely distributed in the rat brain, and are especially abundant in the limbic system. They are involved in the basal cholinergic neurons in rats. The involvement of basal forebrain-septal cholinergic neurons in learning and memory has been well documented.

The cholinergic neuronal differentiation factor (CNDF) is a cytokine that has a wide-range of activities in embryonic stem cells, hepatocytes, adipocytes, neurons and some hemopoietic cell lines. In the nervous system, CNDF can support motor and sensory neuron survival [2], switch autonomic nerve signaling from noradrenergic to cholinergic type and regulate neuropeptide expression in cultured sympathetic neurons. Furthermore, CNDF is involved in responses to damage of the central as well as peripheral nervous systems [3]. Although CNDF mRNA was detected in several brain regions such as cortex, hippocampus and hypothalamus [4], its role in the central nervous system is little known.

Since the distribution of CNDF mRNA in the rat brain is largely overlapped by NGF and BDNF mRNA, we think the expression of CNDF may also be influenced by the administration of AVP(4-8). The purpose of our investigation is to understand the mechanism of action of AVP(4-8) involved in changes of gene expression.

Results and Discussion

We used the semi-quantitative reverse transcription (RT)-PCR method to examine the CNDF mRNA levels in rat brain. GAPDH cDNA was co-amplified as an internal control. We first confirmed the specificity of the amplified band by Southern blot. Then we chose the PCR cycles to ensure that both CNDF and GAPDH were assayed in the exponential range of amplification. Because CNDF cDNA was exponentially amplified from 27 to 36 cycles, we amplified CNDF cDNA for 33 cycles in the subsequent assay under the same condition. Similarly, the GAPDH cDNA was amplified for 24 cycles in the following PCR (data not shown).

We then examined the time course of CNDF mRNA expression after administration of AVP(4-8). The rats were subcutaneously injected with AVP(4-8) at 1.7μg/kg body weight and subsequently decapitated. The total mRNA was extracted separately from the hippocampus and cortex, reverse-transcribed into cDNA and then amplified with incorporated ^{32}P. After agarose gel separation, the amplified DNA band was autoradiographed, excised and counted by Cerenkov counter. As shown in Fig. 1, CNDF mRNA expression in the hippocampus was increased with the time of administration. At 12 hours after administration, the mRNA was significantly increased to 2.62 times compared with the control ($P<0.001$). The expression peak was 3.02 times that of control at hour 18. The time profile of CNDF mRNA expression in the rat cortex was the same as in the hippocampus, but the intensity of enhancement was greater. The expression peak in the cortex at hour 18 was 5.33 times that of control ($P<0.001$).

We then compared the influence of AVP(4-8) analogues on CNDF mRNA expression in rat brain. The rats were treated with different peptides and the total RNA was prepared after 18 hours. As shown in Table 1, RT-PCR assay revealed that NLPR, which is an agonist of AVP(4-8) receptor, enhanced the CNDF mRNA transcription to 5.76 times that of control. From the behavioral experiments we know that NLPR is more potent than AVP(4-8) in facilitating learning and memory in rats. Here we see a similar result in enhancing CNDF transcription. ZDC(C)PR, which from our previous reports is an antagonist of AVP(4-8), had little effect on CNDF transcription. While co-administered with AVP(4-8), it decreased the transcription intensity of CNDF enhanced by AVP(4-8) from 3.25 to 1.86 folds. AVP increased CNDF mRNA expression less than AVP(4-8), while oxytocin, which has no behavioral activity, also had no effect on CNDF transcription.

Comparing to our previous results, AVP(4-8) and its analogues induced similar increases of NGF and BDNF mRNA transcription in rat brain. However, NGF and BDNF mRNA expression peaked at hour 12. The mRNA of CNDF, NGF and BDNF are allabundant in the limbic system, especially in the hippocampus and cortex. Their expressions were all enhanced by the administration of AVP(4-8). It is postulated that CNDF played a role similar to NGF and BDNF in the mechanism of AVP(4-8) enhanced learning and memory. These results also suggest that the CNDF gene may be one of the targets of AVP(4-8). Since many cholinergic fibers terminate in the hippocampus region, it is possible that the expression of the CNDF gene affects cholinergic neuronal function by a neuropoietic role and cholinergic phenotypic maintenance.

In conclusion, using reverse transcription-PCR, we examined the effect of AVP(4-8) and its analogues on the expression of CNDF mRNA in rat brain. We found that the mRNA transcription of CNDF in rat brain was enhanced and peaked at hour 18 by administration of AVP(4-8). The effects of AVP(4-8) and its analogues to enhance the expression of CNDF were parallel to their activities in facilitating the rat learning and memory.

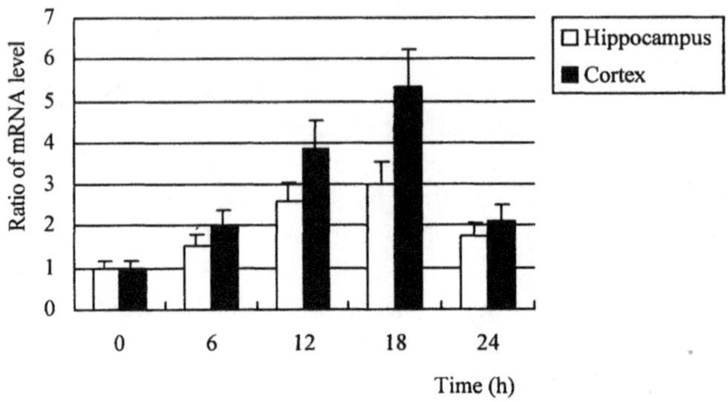

Fig. 1. Time course of CNDF mRNA expression after administration of AVP(4-8). Data were expressed as the ratio of CNDF cDNA to the control (Time 0) after being standardized with GAPDH cDNA (n=3).

*Table 1. Influence of AVP(4-8) and its analogues on CNDF mRNA expression. Data were expressed as the ratio of CNDF cDNA to the control (Time 0) after being standardized with GAPDH cDNA (mean±s.e., n=3). *P<0.01, compared to control; **P<0.01, compared to ZNC(C)PR group.*

	Hippocampus	Cortex
Control	1.00 ± 0.00	1.00 ± 0.00
ZNC(C)PR	3.25 ± 0.54*	3.98 ± 0.63*
NLPR	5.76 ± 0.73*	5.62 ± 0.71*
AVP	2.04 ± 0.64*	2.10 ± 0.48*
ZDC(C)PR	1.06 ± 0.51	1.18 ± 0.56
ZNC(C)PR+ZDC(C)PR	1.86 ± 0.55**	1.95 ± 0.60**
OXT	1.14 ± 0.48	0.92 ± 0.49

References

1. Zhou, A.W., Li, W.X., Guo, J. and Du, Y.C., Peptides, 18 (1997) 1179.
2. Fann, M.J. and Patterson, P.H., Proc. Natl. Acad. Sci. USA, 91 (1994) 43.
3. Banner, L.R. and Patterson, P.H., Proc. Natl. Acad. Sci. USA, 91 (1994) 7109.
4. Minami, M., Kuraishi, Y. and Satoh, M., Biochem. Biophys. Res. Comm., 76 (1991) 593.

The synthesis and opioid activities of nociceptin and its fragments

Shou-Liang Dong, Rui Wang, Tao Wang and Xiao-Yu Hu
Department of Biochemistry and Molecular Biology, School of Life Science, Lanzhou University, Lanzhou, 730000, China

Introduction

A novel neuropeptide named Nociceptin or Orphanin FQ (henceforth called NC) was identified in the rat and porcine brain at the end of 1995 [1, 2]. It is the endogenous ligand of an orphan receptor (LC132 or ORL1, abbreviated ORL1) [3], a new member in the opioid receptor family. This heptadecapeptide has a sequence of Phe-Gly-Gly-Phe-Thr-Gly-Ala-Arg-Lys-Ser-Ala-Arg-Lys-Leu-Ala-Asn-Gln-OH, and shows structural similarities with the mammalian opioid peptide dynorphin A, except for Phe^1 and a different distribution of the basic residues. Its receptor does not bind any of the known opioid ligands with satisfactory affinity and is distinct from the classical opioid receptor in distribution (abundant in the central nervous system, CNS). NC is distinct from other endogenous opioid peptide in pain modulation. It elicits hyperalgesia and has also been found to reverse opioid-mediated analgesia in the mouse. When applied *in vitro*, NC has been shown to inhibit the electrically evoked contractions in mouse vas deferens (MVD) acting through the activation of the ORL1 receptor [4]. It therefore appears that, despite structural similarities with dynorphin A and the identity of some biological effects (e. g. inhibition of sympathetic as well as inhibition of the forskolin-induced accumulation of cAMP), the functional sites of opioids and that of nociceptin are pharmacologically different.

In the present study, we tried to explore the functional role of the nociceptin molecule and identify the message and address sequences that enter into its receptor interaction and immune activity. For this purpose, NC and four of its fragments were synthesized and MVD assay, nociceptive activity were tested.

Results and Discution

MVD assay

The mouse vas deferens is a two receptor system that responds to both the ligand of δ receptor and that of ORL1 receptor with a concentration-dependent reduction of electrically induced twitches. The results are shown in the Table 1. The effect of a δ receptor agonist is found to be reversed by the nonselective opioid receptor antagonist naloxone, where as Naloxone is completely inactive (when applied in a concentration of 1 μM) against the effect of an ORL1 receptor agonist such as nociceptin [4]. We were able to test the changes of the fragment of NC in receptor selectivity. NC, NC(1-15)NH_2 and NC(1-13)NH_2 have the same activity and receptor selectivity. The receptor selectivity of NC(1-11)NH_2 is altered. Because naloxone could reverse its effect about 48.6%, NC(1-11)NH_2 could bind not only ORL1 receptor but also□receptor. NC(1-5)NH_2 binds δ receptor almost completely.

Table 1. Primary structures, abbreviated names and the IC_{50} of MVD assay.

Abbreviated names	Primary structures	IC_{50} of MVD	Naloxone at presence
NC	FGGFTGARKSARKLANQ-OH	20.3	No
NC(1-15)NH_2	FGGFTGARKSARKLA-NH_2	19.0	No
NC(1-13)NH_2	FGGFTGARKSARK-NH_2	19.2	No
NC(1-11)NH_2	FGGFTGARKSA-NH_2	>50000	46.8%
NC(1-5)NH_2	FGGFT-NH_2	>50000	100%

Pain modulation

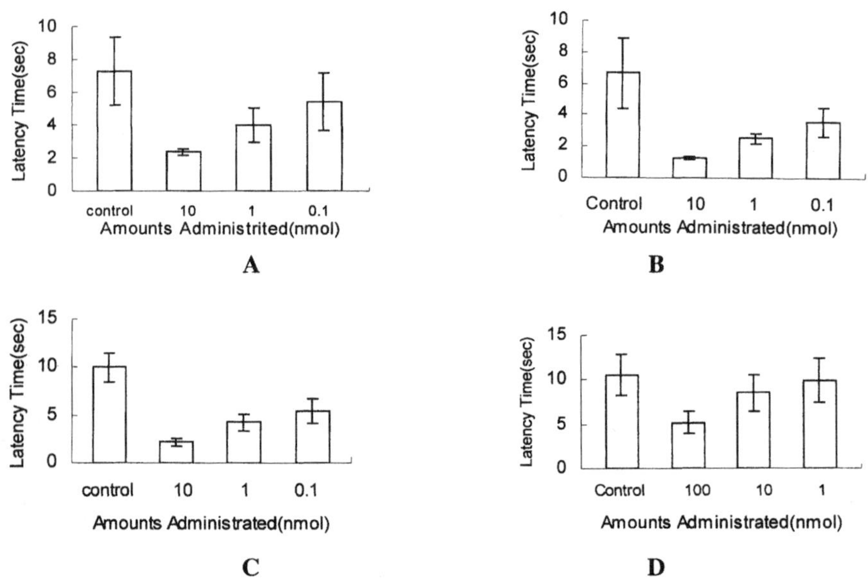

Fig. 1. Hyperalgesia effect of NC and its fragments. ***A***: NC; ***B***: NC(1-15)NH_2; ***C***: NC(1-13)NH_2; ***D***: NC(1-5)NH_2

NC and its fragments were evaluated for the in vivo activity in mice after 10 nmol, the reaction time was reduced about 80%. NC(1-11)NH_2 and NC(1-5)NH_2 intracerebroventricular administration (i.c.v.). NC, NC(1-15)NH_2 and NC(1-13)NH_2 elicited hyperalgesia with 1 nmol to 10 nmol (i.c.v. per mouse). At the dose of elicited hyperalgesia with 10 nmol to 100 nmol (i.c.v. per mouse).

Acknowledgements

This work was supported by grants from the National Natural Science Foundation of China, Fok Ying Tung Education Foundation and the State Education Commission of China.

References

1. Mollereau, C., Parmentier, M. and Mailleux, P., FEBS Letter, 341 (1994) 33.
2. Meunier, J.C., Mollereau, C. and Toll L., Nature, 377 (1995) 532.
3. Reinscheid, R.K., Nothacker, H.P. and Bourson, A., Science, 270 (1995) 792.
4. Berzetei-Gurske, I.P., Schwartz, R.W. and Toll, L., Eur. J. Pharmacol., 302 (1996) R1.

The effect of Glu position change on the opioid activity of deltorphin II

Rui Wang, Ding-Jian Yang, Ya-Ping Ma, Xiao-Feng Huo, Jian-Min Shen and Xiao-Yu Hu
Department of Biochemistry and Molecular Biology, School of Life Science, Lanzhou University, Lanzhou, 730000, China

Introduction

Deltorphin II (Tyr-D-Ala-Phe-Glu-Val-Val-Gly-NH_2, Del II) is a highly selective agonist of δ_{II} opioid receptor [1], which was isolated from the frog *Phyllmedusa Sauvagei* in 1989 [2]. DeltorphinII and dermorphin (Tyr-D-Ala-Phe-Gly-Tyr-Pro-Ser-NH_2, DRM), which is the μ-receptor selective agonist [3], share common generalized N-terminal tripeptide sequences, H-Tyr^1-D-Ala^2-Phe^3, which comprise the message domains of these peptides. The critical structural differences between these two peptides lie in the C-terminal tetrapeptide region, the address domain, adjacent to the N-terminal message domain. Deltorphin II analogs were selected to test the hypothesis of Schwyzer [4] that flexible peptide hormones contain "synchnologic organization", i.e., the polypeptide sequence is composed of proximal regions which exist as distinct message and address domains. The message domain provides information for signal transduction leading to a biological response, while the address domain primarily influences binding affinities and accommodates the elements of selectivity without necessarily affecting transduction. In order to expand our understanding of the characteristics of the C-terminal tetrapeptide address domain of deltorphin II on opioid receptor affinities and selectivities, studies were conducted with several analogs of the peptide in which the spatial distribution of the negatively charged residues was repositioned relative to the hydrophobic residues. Its position 4 is a negative charged amino acid Glu. To study the effect of Glu on the function of deltorphin II, we designed and synthesized a series of analogs of deltorphin II by SPPS (solid phase peptide synthesis), they were: (fig. 1).

[Val^4, Glu^5]Del II, [Val^4, Glu^6]Del II, [Gly^4, Glu^7]Del II

NH_2-Tyr-D-Ala-Phe-***Glu***-Val-Val-Gly-NH_2 [Glu^4]DELII
NH_2-Tyr-D-Ala-Phe-Val-***Glu***-Val-Gly-NH_2 [Val^4, Glu^5]Del II
NH_2-Tyr-D-Ala-Phe-Val-Val-***Glu***-Gly-NH_2 [Val^4, Glu^6]Del II
NH_2-Tyr-D-Ala-Phe-Gly-Val-Val-***Glu***-NH_2 [Gly^4, Glu^7]Del II

Fig. 1. The sequence of deltorphin analogues.

Results and Discussion

Their opioid activity was evaluated in the mouse vas deferens (MVD) and guinea pig ileum (GPI) assays (*in vitro*) (table 1).

Table 1. Mouse vas deferens (MVD) and guinea pig ileum (GPI) Assays of deltorphin II and its analogs.

Compounds	IC_{50}(nM) MVD	GPI	Sel. Rat. GPI/MVD
Del II	0.53 ±0.03	3200 ±551	6037.7
[Val^4, Glu^5]Del II	3.75 ±0.20	8886 ±947	2369.6
[Val^4, Glu^6]Del II	13.29 ±0.57	9266 ±1020	697.2
[Gly^4, Glu^7]Del II	2.26 ±0.14	823 ±126	368.1

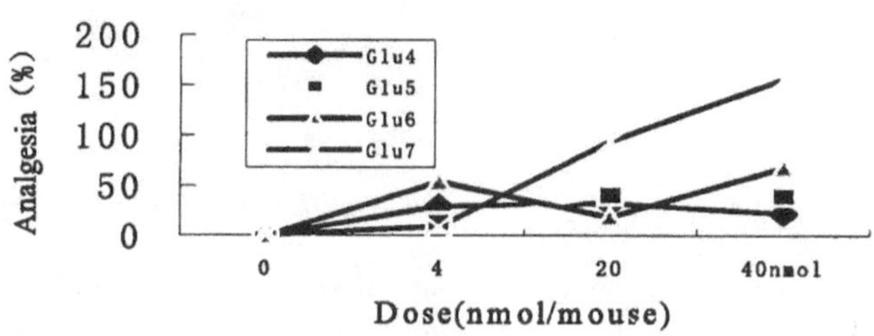

Fig. 2. Dose-analgesia effect lines of the analogs (i.c.v.).

The results showed that the δ-affinity was Del II>[Gly^4, Glu^7]Del II>[Val^4, Glu^5]Del II>[Val^4,Glu^6]Del II ; the μ-affinity was [Gly^4,Glu^7]Del II>[Val^4, Glu^6]Del II>[Val^4,Glu^5]Del II> Del II; the δ-selectivity was Del II>[Val^4, Glu^5]Del II>[Val^4, Glu^6]Del II>[Gly^4, Glu^7]Del II. The antinociceptive activity was [Gly^4, Glu^7]Del II> Del II>[Val^4,Glu^6]Del II>[Val^4,Glu^5]Del II after *i.c.v. (introcerebroventricular).*(fig. 2) The activity of anti-Ang II of [Gly^4,Glu^7]Del II on blood pressure and cholecyst was higher than that of Del II (fig. 3).

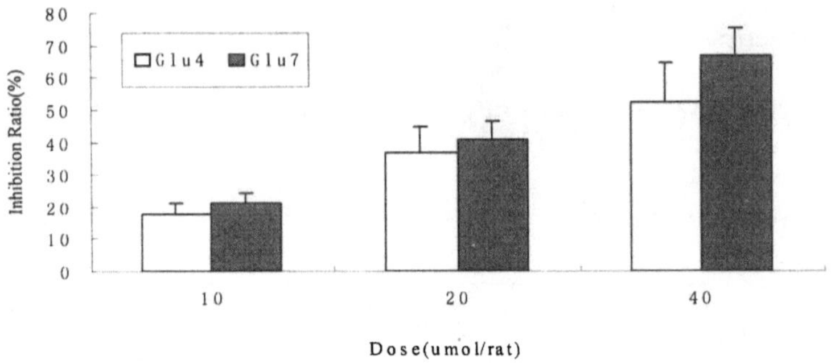

Fig. 3.1 The activity of anti-Ang II of [Gly⁴, Glu⁷]Del II and Del II on blood pressure.

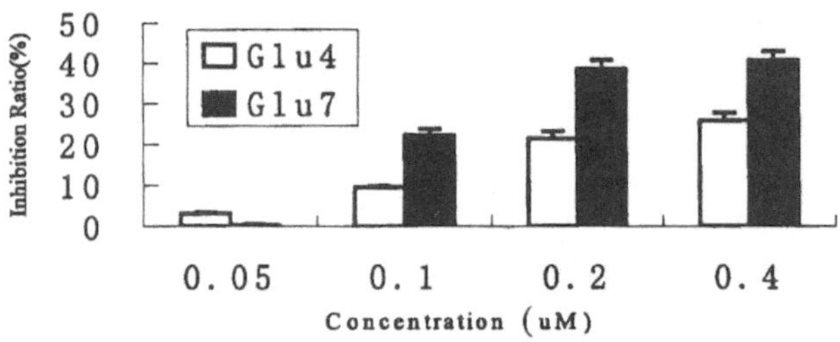

Figure 3.2 The activity of anti-Ang II of [Gly⁴, Glu⁷]Del II and Del II on cholecyst contraction.

The results showed that an anionic group in the address domain does not appear essential for high δ-affinity, but is required for the remarkable δ-selectivities of deltorphin II. Repositioning the Glu residue within the address domain of deltorphin II modifies binding affinity and selectivity for both opioid receptor types. It was concluded that the locations of charged groups (glutamic acid) relative to hydrophobic residues in the address domain of deltorphin II were critical determinants for both δ-affinity and δ-selectivity.

Acknowledgements

This work was supported by grants from the National Natural Science Foundation of China (No: 29442010, No: 29502007), Fok Ying Tung Education Foundation and the State Education Commission of China.

References

1. Hruby, V.J., Life Sci., 50 (1992) PL75.
2. Erspamer, V., Melchiorri, P., Erspamer, G. F., Negri, L., Corsi, R., Severini, C., Barra, D., Sinmaco, M. and Kreil, G., Proc. Natl. Acad. Sci., 86 (1989) 5188.

3. (a) Montecucchi, P.C., de Castiglione, R., Piani, S., Gozzini, L. and Erspamer, V., Int. J. Pept. Prot. Res., 17 (1981275.
   (b) Montecucchi, P.C., de Castiglione, R. and Erspamer, V., Int. J. Pept. Pr) otein Res., 17 (1981) 316.
4. Schwyzer, R., Ann. N. Y. Acad. Sci., 3 (1977) 297.

Isolation and purification of a series of peptides from *Panax notoginseng*

Jian-Guo Ji, Yun-Hua Ye, Ya-Wei Zhou and Qi-Yi Xing
Department of Chemistry, Peking University, Beijing, 100871, China

Introduction

Panax notoginseng is a famous traditional Chinese medicine used for thousands of years for the treatment of bleeding, cerebravascular diseases, coronary heart disease, etc [1]. Modern research on *Panax notoginseng* began in 1937 by Chou T. Q. and Zhu T. H. [2]; they separated the notoginseng saponins from the root of *Panax notoginseng* for the first time. The present study revealed that notoginseng contains volatile oils, sapanins, flavones, sterols, saccharides, proteins, amino acids and minor elements. However, few researchers focused on the study of water-soluble constituents, especially on the peptide constituents in boiling water extracts. According to the traditional refining method of the *Panax* species in Chinese medicine, we focus our study on the nonprotein amino acids and peptides in boiling water extracts and evaluated [3-7]. This paper reports the results of the isolation and purification which verified that a series of peptides exist in the boiling water extracts of *Panax notoginseng*.

Results and Discussions

Panax notoginseng roots were collected from Yunnan province of southwestern China. The dry main root (400 g) of *Panax notoginseng* was smashed into powder and passed through 100 meshes. The powder was extracted with water under boiling for 2 hours, the mixture was filtered, and the residue was extracted another time. The filtrate was combined and purified with ion exchange resin, Sephadex G-15 and RP-HPLC. More than ten constituents were collected after HPLC purification. These constituents were composed of different amino acids after 6N HCl hydrolysis, although there were no free amino acids in the analysis before 6N HCl hydrolysis. The peptide components and their amino acid composition are shown in table 1. Among them, YN-3H11 was composed of Glu, Gly and Cys. The amino acid ratio of YN-3H11 was Glu : Gly : Cys = 1 : 0.9 : 0.54 and its N-terminal was Glu analysed by 491 protein sequencer; its C-terminal was -Gly-OH by carboxylpeptidase A degradation method. The FAB-MS and MALDI-TOF MS did not give its molecular weight, and the RP-HPLC retention time was different from that of the oxidized glutathione and glutathione. (Retention time: YN-3H11: 10.94 minutes; the oxidized glutathione: 11.91 minutes; glutathione: 10.13 minutes). Reversed Phase High Perfomance Liquid Chromatography (RP-HPLC) was performed under the following conditions: Waters 600E, W-Porex 5C18, 250 × 10 mm (Phenomenex), Eluent: 2 ml/min., 1% acetonitrile and 0.1% trifluoroacetic acid (TFA) to 19% acetonitrile and 0.1% trifluoroacetic acid (TFA), detection: 220 nm. The determination of their sequences and studies on their bioactivities are in progress.

Table 1. The peptide constituents and their amino acids composition.

Peptide constituents	Amino acid composition after 6N HCl hydrolysis
YN-3(6)2e2	Asp, Glu, Gly, Cys
YN-3(6)2e3	Glu, Gly, Arg
Yn-3H1	Glu, Gly, Cys
YN-3B2	Glu, Gly, Arg
YN-3J1	Asp, Glu, Gly, Tyr
YN-3H3	Asp, Glu, Gly
YN-3J2	Asp, Thr, Glu, Gly, Tyr
YN-3H2	Asp, Glu, Gly, Cys, Tyr
YN-3H4	Asp, Glu, Gly, Cys, Tyr, Phe
YN-3E3	Asp, Ser, Thr, Glu, Pro, Gly, Ala, Val, Ile, Leu, Lys, Arg
YN-3E4	Asp, Ser, Thr, Glu, Pro, Gly, Ala, Val, Ile, Leu, Lys, Arg
YN-3E5	Asp, Ser, Thr, Glu, Pro, Gly, Ala, Val, Met, Ile, Leu, Tyr, Phe, Lys, Arg
YN-3E6	Asp, Ser, Thr, Glu, Pro, Gly, Ala, Val, Met, Ile, Leu, Tyr, Phe, Lys, Arg

Acknowledgements

This work was supported by the Doctoral Program Foundation of Institution of Higher Education.

References

1. Li, Q., Ye, Y.H. and Xing, Q.Y., Chem. J. Chin. Univ., 17 (1996) 1886.
2. Chou, T.Q. and Chu, T.H., China J. Physiology, 12 (1937) 59.
3. Yang, L., Ye, Y.H. and Xing, Q.Y., Chinese Chemical Letters, 1(1990) 51.
4. Ye, Y.H., Long, Y.C. and Xing, Q.Y., Proceedings of '92 International Ginseng Symposium, Changchun, China, 1992, p.98.
5. Xing, Q.Y., Ye, Y.H. and Yang, L., Proceedings of the 6th International Ginseng Symposium, Seoul, Korea, 1993, p.124.
6. Chen, Z.K., Ye, Y.H. and Xing, Q.Y., Chinese Chemical Letters, 7 (1996) 337.
7. Long, Y.C., Ye, Y.H. and Xing, Q.Y., Int. J. Peptide Protein Res., 47 (1996) 42.

Study of the relationship between structure and function of HIV-1 gp 41 N terminus fusion peptide

Man Wu [a], Song-Qing Nie [a], Yang Qiu [a], Ke-Chun Lin [a], Shao-Xiong Wang [b] and Sen-Fang Sui [b]

[a.] *Department of Biophysics, Beijing Medical University, Beijing, 100083;* [b.] *Department of Biological Science and Technology, Tsinghua University, Beijing, 100084, China*

Introduction

The human immunodeficiency virus type 1(HIV-1) envelope glycoprotein consists of two noncovalently associated subunits, gp120 and gp41, which are generated by proteolytic cleavage of a precursor polypeptide, gp160. This cleavage generates in gp41 the exposure of a highly hydrophobic N-terminal amino acid stretch, which has been proposed to interact with the lipid membrane during fusion events. The mode of interaction of the fusion peptide with the lipid membrane, in particular the mechanism of insertion into the membrane and the mechanism by which it facilitates fusion, is not fully understood [1]. In an attempt to establish a relationship between peptide structure and function, we have investigated in a liposomal system the activity of a 23-residue synthetic peptide corresponding to the N-terminal extremity of HIV-1 gp41 and its mutant.

Results and Discussion

The peptides used in the experiment are a 23-residue synthetic peptide (HIV_{wt}) representing HIV-1 gp41 N terminus and its mutant V_2E (HIV_{Glu}, a single polar substitution V→ E at position 2).

By fluorescence study, HIV_{wt} bound to negative phospholipid stronger than to lipids devoid of a net charge; the ability of HIV_{Glu} to bind with lipid was less than the wild type. In order to investigate the insertion further, fluorescence quenching test was used [2]. This experiment showed that HIV_{wt} inserted deeply into negative phospholipid (fig.1); HIV_{Glu} did not insert into lipids although it did bind with lipids.The molecular dynamics revealed by fluorescence polarization showed that membrane fluidity is increased after interaction with HIV_{wt}. It suggests HIV_{wt} destablizes membrane more than HIV_{Glu}.

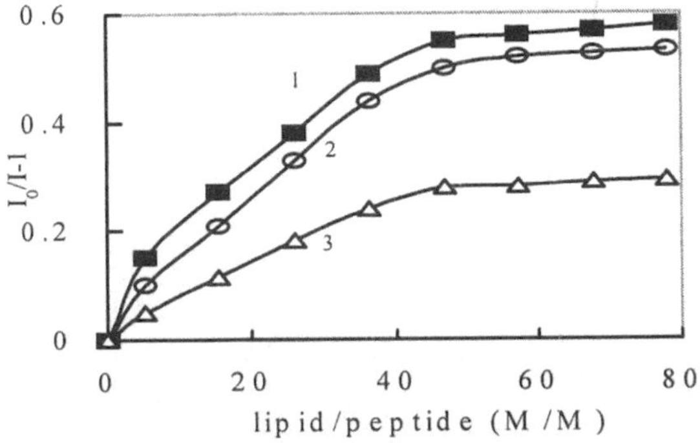

Fig. 1. Penetration of HIV_{wt} into lipid vesicles using spin-labeled fatty acids as quenchers. Increasing fluorescence was recorded. 5% probe (M/M). (1) 12-DSA-POPG ,(2) 5-DSA-POPG, (3) 9-DSA-POPG.

A monolayer study confirmed the above results. The insertion ability of HIV_{wt} into negative charged lipids is significantly higher than into neutral ones. In the case of POPG monolayer, the critical insertion pressure of HIV_{wt} was 43 mN/m but for HIV_{Glu} was only 31mN/m (fig. 2 and 3). This suggests that the insertion of HIV_{wt} into phospholipid monolayers was driven not only by electrostatic force but also by hydrophobic interactions.

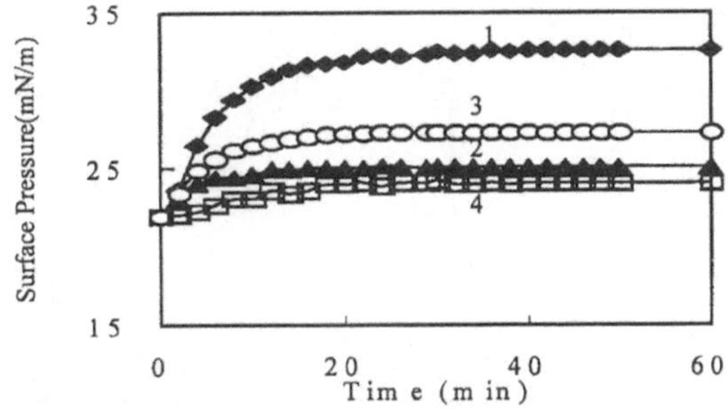

Fig. 2. The time course of insertion of HIV_{wt} into POPG (1), POPC (2), HIV_{Glu} into POPG (3) and POPC (4). The initial pressure is 22 mM/N. Subphase is Hepes buffer. T. 22 °C.

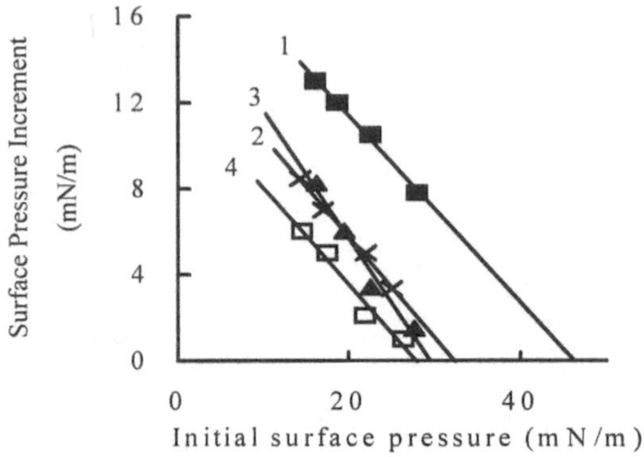

Fig. 3. The relationship between the pressure increment and the initial pressure in the interaction of HIV_{wt} with POPG(1),POPC(2) and HIV_{Glu} with POPG(3), POPC (4).

A study of Infrared spectroscopy showed that at high peptide: lipid ratios, anti-parallel β-sheet conformation was observed for HIV_{wt}; at lower peptide: lipid ratios, β-structure partly converted into α-helices (table 1) [3].

Table 1. Quantitative studies of the secondary structure of HIV_{wt} in complexes with POPG vesicles by FTIR .

	HIV_{wt}		POPG/HIV_{wt}(60:1)		POPG/HIV_{wt}(150:1)	
	Bands (cm^{-1})	Area (%)	Bands (cm^{-1})	Area (%)	Bands (cm^{-1})	Area (%)
α-helix	1659.3	4.7	1650.8	16.1	1650.9	24.3
β-sheet	1629.7 1621.2	68.1	1685.5 1682 1624.6	63.5	1683.6 1635.8	39.8
β-turns	1678.5 1665.9	17.2	1668.8	12.9	1671.2 1667.4	7.0
random coils	1644.8 1641.2	14.3	—	—	—	—

In summary, HIV_{wt} can insert into acyl chains of membranes and thus it is easy to facilitate fusion. HIV_{Glu}, however, may spread parallel to the plane of the membrane, and cannot induce membrane fusion. The conformation and orientation of these two peptides should to be studied further.

Acknowledgements

We acknowledge the financial support obtained from National Natural Science Foundation of China, also Dr. Jan Wilschut who sent to us two peptides.

References

1. Martin, T. J., Virol., 70 (1996) 298.
2. Wang.Q.D., Biochem. Biophys. Acta., 1324 (1997) 69.
3. Pereira, F., Biophys. J., 73 (1997) 1977.[asd1]

Hypotensive activities of nociceptin and its fragments

Rui Wang, Tao Wang, Shou-Liang Dong, Qiang Chen and Xiao-Yu Hu

Department of Biochemistry and Molecular Biology, School of Life Science, Lanzhou University, Lanzhou, 730000, China

Introduction

Nociceptin, novel neuropeptide (H_2N-Phe-Gly-Gly-Phe-Thr-Gly-Ala-Arg-Lys-Ser-Ala-Arg-Lys-Leu-Ala-Asn-Gln-COOH) also know as orphanin FQ is a endogenous ligand for the "orphan" opioid receptor ORL1 [1, 2]. Nociceptin is a 17 amino acid peptide that shares structural homology with the dynorphin family of peptides. Nociceptin differs from other opioid peptides in that it does not possess the N-terminal tyrosine residue that is essential for activity at μ, δ, and κ opioid receptors. The isolation of nociceptin was based on the ability of the peptide to inhibit adenylyl cyclase in ORL1 transfected cells. Although this signaling mechanism is similar to other opioid agonists when their receptors are activated, nociceptin induces hyperalgesia when injected into the cerebral ventricles of the rat. In addition, nociceptin has potent diuretic and natriuretic activity in the rat [3].

It has been suggested that opioid receptors have a common binding site that interacts with the N-terminal pentapeptide moiety (Tyr-Gly-Gly-Phe-Met/Leu) and that the C-terminal amino acid determines the degree of relative selectivity for the ligand's respective receptor subtype [4]. It has been suggested, therefore, that the Phe in the first position of the nociceptin sequence conveys the selectivity of this novel ligand for ORL1 receptor [4]. Champion and Kadowitz have shown that nociceptin and its analog, [Tyr^1]-nociceptin, have novel hypotensive activity in rats. However, the effects C-terminal sequence of nociceptin on vasodepressor activity is not known. Therefore, nociceptin (NC) and its four fragments, NC(1-15)NH_2, NC(1-13)NH_2, NC(1-11)NH_2, NC(1-5)NH_2, were prepared and investigated in the rat systemic vascular bed of rats, in an attempt to identify the sequence involved in the activation (message) and in the binding (address) of nociceptin to its receptor.

Results and Discussion

The effects of NC and its four fragments on systemic arterial pressure in the rats were investigated (table 1.).

The injection of NC, NC(1-15)NH_2, NC(1-13)NH_2, into the jugular vein in a dose of 30 nmol/kg caused a marked decrease in systemic arterial pressure The response was reproducible with respect to time, with systemic arterial pressure slowly returning to baseline over a 6-9 min period. The decreases in systemic arterial pressure in response to NC, NC(1-15)NH_2, NC(1-13)NH_2, were dose-related when the nociceptin fragments was injected in dose of 1-50 nmol/kg i.v. (fig. 1). Decreases in systemic arterial pressure in response to NC, NC(1-15)NH_2, NC(1-13)NH_2, were similar. NC(1-11)NH_2, NC(1-5)NH_2 injected in a dose of 100 nmol/kg had no hypotensive activity .

Table 1. Primary structures, abbreviated names.

Abbreviated names	Primary structures
NC	F-G-G-F-T-G-A-R-K-S-A-R-K-L-A-N-Q-OH
NC(1-15)NH2	F-G-G-F-T-G-A-R-K-S-A-R-K-L-A-NH2
NC(1-13)NH2	F-G-G-F-T-G-A-R-K-S-A-R-K-NH2
NC(1-11)NH2	F-G-G-F-T-G-A-R-K-S-A-NH2
NC(1-5)NH2	F-G-G-F-T-NH2

The effects of the opioid receptor antagonist naloxone on decreases in systemic arterial pressure in response to nociceptin and its fragments were investigated, and these data are summarized in Fig.2. Decreases in systemic arterial pressure in response to NC, NC(1-15)NH_2, NC(1-13)NH_2, NC(1-5)NH_2, did not change following administration of naloxone in a dose of 2 mg/kg i.v. (fig. 3). However, NC(1-11)NH_2 showed hypotensive activity in the presence of naloxone (at 100 nmol/kg i.v. with systemic arterial pressure decreased 3 kpa).

Results of the present study show that decreases in systemic arterial pressure in response to NC, NC(1-15)NH_2 and NC(1-13)NH_2, are not altered by naloxone. The data suggest that such decreases are not mediated by a naloxone-sensitive mechanism. Therefore, NC, NC(1-15)NH_2 and NC(1-13)NH_2 may bind only to ORL1. Moreover, because NC(1-15)NH_2 and NC(1-13)NH_2 were comparable in potency to nociceptin, the C-terminal tetrapeptide is not critical for nociceptin binding to its receptor. However, NC(1-11)NH_2, without Arg12-Lys13 at its C-terminal, had much lower activity than NC, and less receptor selectivity. It binds not only ORL1 but also δ opioid receptor. NC(1-5)NH_2, had lost almost all the activity of nociceptin because it lacked the main part of C-terminal.

In summary, the results of the present study show that vasodepressor responses to nociceptin and its fragments are not mediated by the activation of a naloxone-sensitive opioid receptor and the entire sequence of NC may not be required for full hypotensive activity since NC(1-13)NH_2 is as active as NC .The degree receptor selectivity was determined by the nociceptin C-terminal residues.

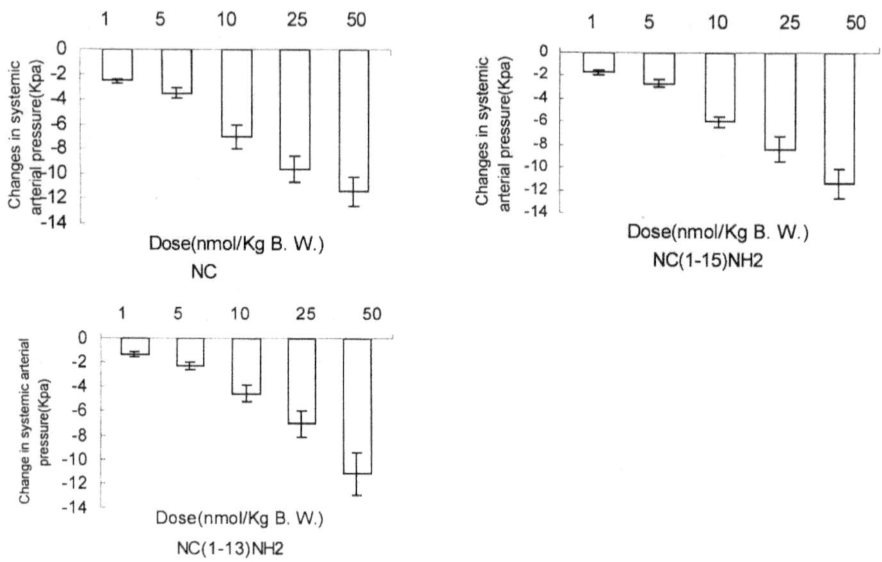

Fig. 1. Bar graphs showing dose-dependent decrease in systemic arterial pressure in response to intravenous injection of NC and its fragments in the rats.

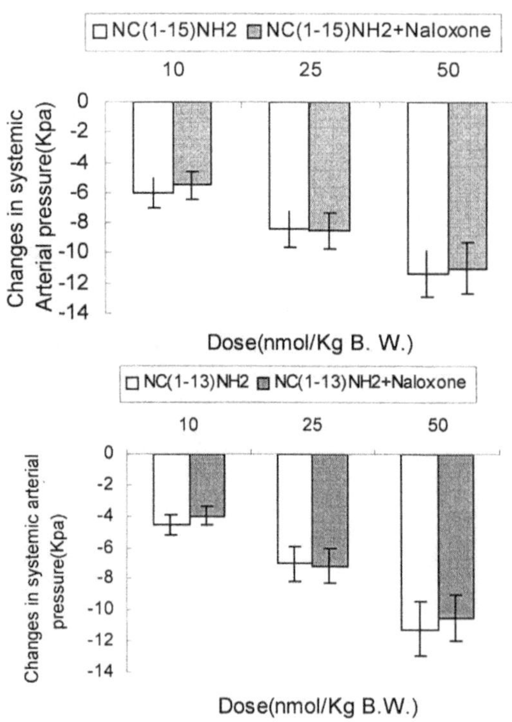

Fig. 2. Bar graphs showing the influence of the opioid receptor antagonist naloxone on response to NC(1-15)NH_2 and NC(1-13)NH_2.

Acknowledgements

This work was supported by grants from the National Natural Science Foundation of China, Fok Ying Tung Education Foundation and the State Education Commission of China.

References

1. Mollereau, C., Parmentier, M. and Mailleux, P., FEBS Letter, 341 (1994) 33.
2. Meunier, J.C., Mollereau, C. and Toll, L., Nature, 377 (1995) 532.
3. Kapusta, D.R., Sezen, S.F. and Chang, J.K., Life Sci., 70 (1997) PL25.
4. Champion, H.C. and Kadowitz, J., Biochem. Biopghys. Res. Comman., 234 (1997) 2.

The design and synthesis of endomorphins and their analogues

Xiao-Feng Huo, Wei-Hua Ren, Ning Wu and Rui Wang
Department of Biochemistry and Molecular Biology, School of Life Science, Lanzhou University, Lanzhou, 730000, China

Introduction

The peptides that are considered to be endogenous agonists for δ (enkephalin) and κ (dynorphin) opiate receptors have been identified in mammalian, but none have been found to have preference for the μ receptor. Because morphine and other compounds that are clinically useful but are opening to abuse act primarily at the receptor, it could be important to study the specificity of peptides to this site [1]. Endomorphin 1 and 2 (fig. 1) are newly isolated, potent and selective μ-opiate receptor endogenous agonists [2]. Endomorphin-1 (H-Tyr-Pro-Trp-Phe-NH_2) has a high affinity (k_i=360 pM) and selectivity (4,000- and 15,000- fold preference over the δ and κ receptors) to the μ receptor. This peptide is more effective than the μ- selective analogue DAMGO *in vitro* and produces potent and prolonged analgesia in mice. Endomorphin-2 (H-Tyr-Pro-Phe-Phe-NH_2) also has a high affinity (k_i=690 pM) and selectivity (7,000- and 13,000- fold preference over the κ and δ receptors) to the μ receptor. The endomorphins have the highest specificity and affinity to the μ receptor among all endogenous substance so far described [2]. In fact, endomorphin-1 would not have been found if Zadina *et al* hadn't changed amino acid at position 4 of Tyr-W-MIF-1 (Tyr-Pro-Phe-Phe-NH_2) [2]. It has been demonstrated that the development of a single compound with mixed μ agonist/δ antagonist properties may have therapeutic potential. The first known example of a mixed μ agonist/ δ antagonist is an opioid tetrapeptide analogue H-Tyr-Tic-Phe-Phe-NH_2 (TIPP-NH_2) that was found to be a moderately potent μ agonist in the GPI assay and a highly potent δ antagonist in the MVD assay [3]. Endomorphin-2 (H-Tyr-Pro-Phe-Phe-NH_2) differs from TIPP-NH_2 by one amino acid at position 2. In a way, this compound is obtained by transforming endomorphin-2 at position 2. The analogues designed based on endomorphins may have therapeutic potential. Besides endormorphins differ from N-terminal tetrapeptide fragment of traditional opioid peptides at position 2, there is dissimilarity at position 3. Moreover, endomorphin-1 differs from endomorphin-2 at position 3. This implies the more messages are included in position, and especially, it may notably affect endomorphins selectivity. So, we intend to confirm the sense of position 3 of endomorphins by changing its space and charge effects. The studies on transforming at position 2 and 3 of endomorphins will aid in their further researches and development.

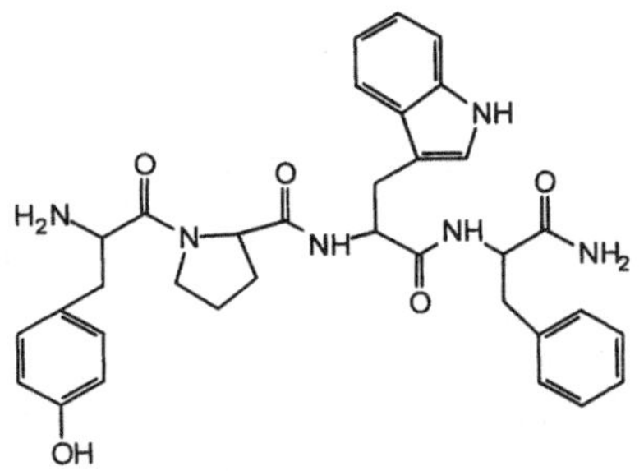

Endomorphin-1

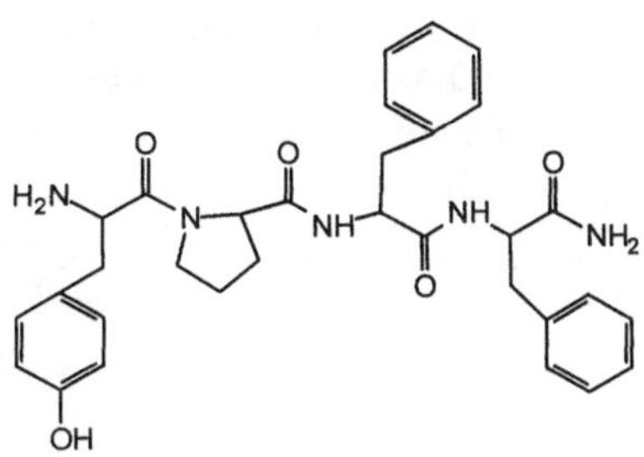

Endomorphin-2

Fig. 1. Schematic representations of endomorphins structure. The amine and phenolic groups of tyrosine and the aromatic group of the third residue (Trp and Phe) are required for the receptor recognition.

Results and Discussion

Based on the work on endomorphins, TIPP-NH_2, and other opioid peptides in the literature, we designed and synthesized seven analogues of endomorphins (table 1).

These peptides used in the study were synthesized step by step by solution-phase procedures according to the sequence from N-terminal to C-terminal. Dicyclohexycarbodiimide (DCC) and N-hydroxybenzotriazole (HOBT) were used as the coupling reagents. Trifluroacetic acid (TFA) was used as the deprotection reagent, and N-methylmorpholin (NMM) was used as the neutrality reagent. The original materials were N-BOC-L-Tyr-OH and unprotected amino acids. The synthesis method was the following four steps.

Table 1. The amino acid sequence of endomorphins and their analogues.

The analogues relative to changing at position 2	The analogues relative to changing at position 3
Tyr-Pro-Trp-Phe-NH_2(Endomorphin-1)	Tyr-Pro-Gly-Phe-NH_2
Tyr-Pro-Phe-Phe-NH_2(Endomorphin-2)	Tyr-Pro-Pro-Phe-NH_2
Tyr-D-Ala-Trp-Phe-NH_2	Tyr-Pro-Tyr-Phe-NH_2
Tyr-D-Pro-Trp-Phe-NH_2	
Tyr-D-Ala-Phe-Phe-NH_2	
Tyr-D-Pro-Phe-Phe-NH_2	

First, the benzyl amino acids were obtained by conventional method to protect the C-terminal [4], and used as amino group component for reserve.

Second, the carboxylic group component whose N-terminal was protected and the amino group component were condensed together.

Third, for next condensation, the carboxylic group was exposed by removing benzyl with saponification and acidification. After such three circulations, the tetrapeptide was gained whose two ends are protected.

Last, using TFA, the BOC group is removed. The OBZL ester is transformed into amide by aminolysis [5], and the object product is obtained. Taking [L-Gly^3]endomorphin as an example, the schematic of the synthesis is shown in scheme 1.

The purity of the synthesized compounds was analyzed by HPLC, and the structure of the products was confirmed by fast atom bombardment mass spectroscopy (FAB-MS) and 400-MHz ^{1}H-NMR spectroscopy. The potency and selectivity of these analogues will be evaluated by radio receptor binding assays and by bioassays using the mouse vas deferens (MVD, δ-receptor assay) and guinea pig ileum (GPI, μ-receptor assay) assays. With a rapid filtration technique, against [^{3}H]-DPDPE (δ-agonist) and [^{3}H]-DAMGO (μ-agonist), binding affinities of the compounds will be measured .The studies on the bioassays and the structure-activity relationships are in the process.

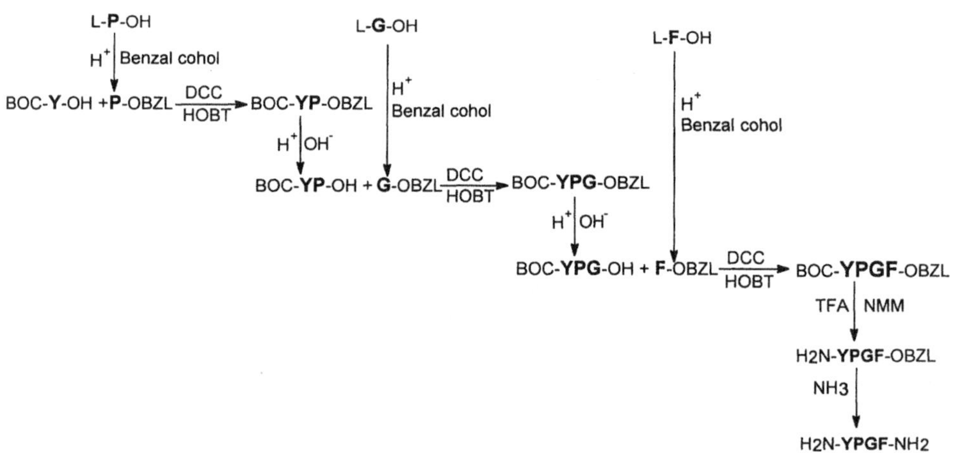

Scheme 1. The schematic of the synthesis of [L-Gly^3]endomorphin.

Acknowledgements

This work was supported by grants from the National Natural Science Foundation of China (No: 29442010, No: 29502007), Fok Ying Tung Education Foundation and the State Education Commission of China .

References

1. Jin, Y.C., Molecular Pharmacology, Tianjin Science and Technology Publishers, Tianjin, 1990, p.444.
2. Zadina, J.E., Hackler, L., Ge, L.J. and Kastin, A.J., Nature, 386 (1997) 499.
3. Schiller, P.W., Schmidt, R., Wilkes, B.C., Weltrowska, G., Nguyen, T.M.D., Chung, N.N. and Lemieux, C., In Lu, G.S., and Tam, J.P. and Du, Y.C. (Eds.) Peptides: Chemistry, Structure and Biology (Proceedings of the 3rd Chinese Peptide Sympoism). ESCOM, Leiden, 1995, p.140.
4. Izumiya, N. and Makisumi, S., Nippon Kagaku Zasshi, 78 (1957) 662.
5. Huang, W.D. and Cheng, C.Q., The Synthesis of Polypeptides, Science Publishers, Shanghai, 1985, p.53-54.

Neuronal apoptosis induced by β-amyloid peptides in vitro

Wen-Xue Li, Ben-Xian Gu and Yu-Cang Du
Shanghai Institute of Biochemistry, Chinese Academy of Sciences, Shanghai, 200031, China

Introduction

Excessive accumulation of the β-amyloid peptide (Aβ) in the brain is characteristic of patients with Alzheimer's disease (AD). The distribution of Aβ correlates well with neuronal degeneration that manifests as neurofibrillary tangles in brain regions that are particularly vulnerable in AD such as the hippocampus [1]. A central issue in the pathophysiology of AD is the role of Aβ in the neurodegeneration process. Exposure of primary hippocampal neurons to Aβ causes neuronal degeneration and intracellular calcium destabilization [2]. Similarly, in vivo injection of Aβ is reported to result in stable β-amyloid deposits accompanied by focal neuronal damage [3]. Several laboratories have reported that the neurotoxicity of Aβ is related to its filamentous or β-sheet aggregate state. However, the mechanism of neurotoxicity induced by Aβ is not fully understood. In this study, we have examined the neurodegenerative effect of Aβ on primary hippocampal neurons in order to provide insight into the question of whether Aβ induces neuronal death via apoptosis.

Results and Discussion

Primary hippocampal neurons were prepared from embryonic day 18 Sprague-Dawley rat pups. After 48 hr culture, the neurons were treated with 25 μM Aβ25-35 for 24 hr. The peptide appeared to be filamentous or flocculent aggregates and adhered to the surface of the neurons. The neurites were obviously cracked into pieces and parts of the neuronal cell bodies were degenerated. In situ TUNEL staining showed that several cell nuclei were labeled positive by terminal deoxynucleotidyl transferase (TdT), suggesting the existence of chromosomal DNA fragments (data not shown). DNA ladder detection also revealed the production of low molecular weight DNA in the neurons treated with Aβ. These results suggested that the Aβ induced neurodegeneration via apoptosis.

We further examined whether cell viability varied with the dose and time of Aβ treatment. After a 48 hr culture the neurons were treated with 5 μM and 25 μM Aβ25-35, respectively. Following continuous culture days, neuron viability was measured using MTT assay. As shown in Fig. 1, 5 μM Aβ25-35 was significantly toxic. After 1d treatment, cell viability was decreased to 85.7%. After 4 days, it was 68.3%. When the concentration of Aβ was increased to 25 μM, the viabilities were 66.3% and 39.7% after 1-day and 4-day treatments, respectively, suggested that the neurotoxicity of Aβ was dose and time-dependent.

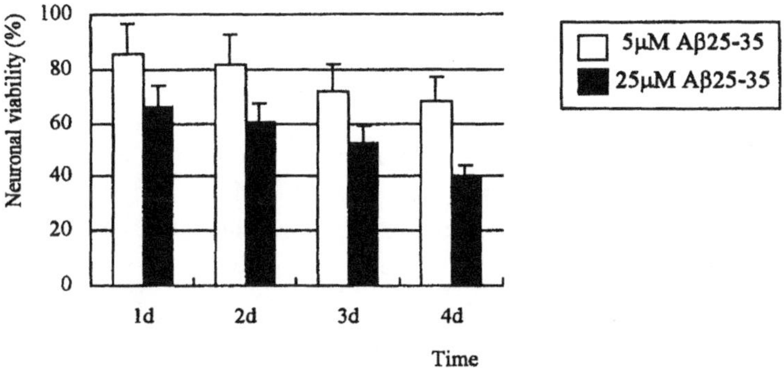

Fig. 1. Cell viability varying with dose and time of Aβ25-35 treatment. Using MTT assay, the cell viability was expressed as the percent of OD_{570nm} between Aβ treated group and control (n=5).

We compared the toxic effects of freshly prepared and aged Aβ1-40. New Aβ1-40 solution contained no or very deposits, while after incubation at 37 °C for 7 days, it became turbid with few fibrillary and flocculent aggregates. As shown in Table 1, after 4-day treatment, 25 μM Aβ1-40 (new) decreased cell viability to 85.0%. When the solution was aged, the corresponding value was 34.7%, suggesting the toxicity of Aβ1-40 was enhanced by aging.

Table 1. Toxicity of different peptides and different solution states. Values were expressed as the percent of OD_{570nm} between Aβ treated group and control (mean±S.E., n=5).

	Aβ25-35	Aβ1-40 (new)	Aβ1-40(aged)	Aβ40-1(aged)
5μM	64.7±3.1	91.0±3.6	60.7±1.5	96.0±3.6
2 5μM	42.0±2.0	85.0±3.0	34.7±2.5	97.0±3.6

We examined the toxicity of Aβ40-1, the sequence-reversed peptide of Aβ1-40. Aged Aβ40-1 solution showed no aggregates, and did not induce degenerative morphological changes. Also it had little effect on cell survival (see Table 1).

In this article, we have evaluated the hypothesis that the β-amyloid peptides induce degeneration of primary hippocampal neurons via apoptosis. The data support our prediction that β-amyloid treated neurons exhibit ribosomal DNA fragmentation, the distinct feature of cell apoptosis. It was reported that exposure of cultured rat hippocampal neurons to β-amyloid peptides caused a rise in cytoplasmic Ca^{2+} and subsequent cell death [2]. Perhaps the destabilization of calcium homeostasis resulted in the neuronal apoptosis. The intracellular signal transduction event inducing neuronal apoptosis needs more research. The neurotoxic potency of Aβ was associated with peptide aggregation, and preaggregation of the peptide ('aging') increased the rapidity and severity of neuronal damage. Other workers have provided evidence supporting the importance of peptide aggregation for manifestation of Aβ neurotoxicity [4, 5]. These findings suggest the biochemical clue for the toxicity of accumulated β-amyloid deposition in AD brain. Deposition of β-amyloid is one of the earliest pathological changes observed in the brains of AD patients. It is possible that the aggregation of Aβ played a key role in the process of neuronal degeneration. However, the signal pathway induced by the assembled Aβ must be further investigated.

References

1. Selkoe, D.J., Annu. Rev. Neurosci., 12 (1989) 463.
2. Mattson, M.P., Tomaselli, K.J. and Rydel, R.E., Brain Res., 621 (1993) 35.
3. Kowall, N.W., Beal, M.F., Busciglio, J., Duffy, J.K. and Yankner, B.A., Proc. Natl. Acad. Sci. USA, 88 (1991) 7247.
4. Pike, C.H., Burdick, D., Walencewicz, A.J., Glabe, C.G. and Cottman, C.W., J. Neurosci., 13 (1993) 1676.
5. Busciglio, J., Lorenzo, A. and Yankner, B.A., Neurobiol. Aging, 13 (1992) 609.

Studies on the spin-labeling technique on deltorphin II and its analogues

Ding-Jian Yang, Rui Wang, Ya-Ping Ma, Yu-Jun Di, Jian-Min Shen and Xiao-Yu Hu

Department of Biochemistry and Molecular Biology, School of Life Science, Lanzhou University, Lanzhou, 730000, China

Introduction

ESR spectrum was in the spin-labeling technique by McConnel *et al* in 1965 [1, 2]. Most experimental work to date has used various forms of nitroxide free radicals, shown schematically as A and B.

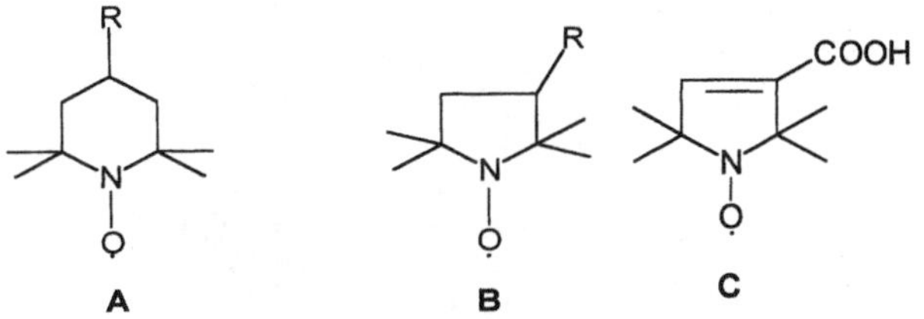

These spin-labels contain a protected nitroxide group, in which the nitrogen atom is bound to two tertiary carbon atoms, and R is a peptide but in general studies it can be another biofunctional group. The free radical nitroxide group then serves as the reporter group reflecting its environment via Electron Resonance Absorption. The changes of ESR spectrum give us information about the microenvironment such as the depth, width, and position of the binding site.

DeltorphinII (Tyr-D-Ala-Phe-Glu-Val-Val-Gly-NH_2, Del II) is a highly selective agonist of δ_{II} opioid receptor [3], which was isolated from the frog *Phyllmedusa Sauvagei* [4]. Its position 4 is a negatively charged amino acid, Glu. We designed and synthesized a series of spin-labeling peptides corresponding to Deltorphin II and its analogues (fig. 1). To study the effect of Glu on the function of deltorphin II and its structure-activity relationships. In addition we tried to further develop the spin-labeling method in peptide, research and study its effect on peptide bioactivities. The labeling material was a stable nitrogen-oxygen free radical (2, 2, 5, 5-tetramethyl-3-carboxyl-3-ene-pyrroline nitrogen-oxygen free radical) that was labeled on the N-terminal by amide bond formation. We studied the opioid bioactivities *in vivo* and *in vitro,* and investigated the changes in the spin-labeling peptides before and after binding to the receptor by ESR spectrum.

NH_2-Tyr-D-Ala-Phe-***<u>Glu</u>***-Val-Val-Gly-NH_2 [Glu^4]DEL II
NH_2-Tyr-D-Ala-Phe-Val-***Glu***-Val-Gly-NH_2 [Val^4,Glu^5]Del
NH_2-Tyr-D-Ala-Phe-Val-Val-***Glu***-Gly-NH_2 [Val^4,Glu^6]Del II
NH_2-Tyr-D-Ala-Phe-Gly-Val-Val-***Glu***-NH_2 [Gly^4,Glu^7]Del II
R□NH-Tyr-D-Ala-Phe-***Glu***-Val-Val-Gly-NH_2 R□[Glu^4]DEL II
R□NH-Tyr-D-Ala-Phe-Val-***Glu***-Val-Gly-NH_2 R□[Val^4,Glu^5]Del
R□NH-Tyr-D-Ala-Phe-Val-Val-***Glu***-Gly-NH_2 R□[Val^4,Glu^6]Del II
R□NH-Tyr-D-Ala-Phe-Gly-Val-Val-***Glu***-NH_2 R□[Gly^4,Glu^7]Del II.

Fig. 1. The sequence of deltorphin II analogues and their spin-labeling derivatives.

Results and Discussion

Their opioid activity was evaluated in the mouse vas deferens (MVD) and guinea pig ileum (GPI) assays (*in vitro*) (table.1).

Table 1. Guinea Pig Ileum (GPI) and Mouse Vas Deferens (MVD) Assay of Opioid Peptide Analogues.

Compound	IC_{50}(nM) MVD	GPI	Sele. ratio GPI/MVD
Del II	0.53 ±0.03	3200 ±551	6037.7
[Val4,Glu5]Del II	3.75 ±0.20	8886 ±947	2369.6
[Val4,Glu6]Del II	13.29 ±0.57	9266 ±1020	697.2
[Gly4,Glu7]Del II	2.26 ±0.14	823 ±126	368.1
R·Del II	0.24 ±0.01	602 ±88	2528.1
R·[Val^4,Glu^5]Del II	0.74 ±0.04	779 ±95	1056.7
R·[Val^4,Glu^6]Del II	8.96 ±0.47	397 ±64	44.4
R·[Gly^4,Glu^7]Del II	0.09 ±0.01	892 ±76	10023.6

The results show that selectivity to δ receptor apparently decreases after spin-labeling, except for [Gly^4, Glu^7]Del II which had 27-fold increased δ selectivity. The affinity to δ receptor had an apparent increase.

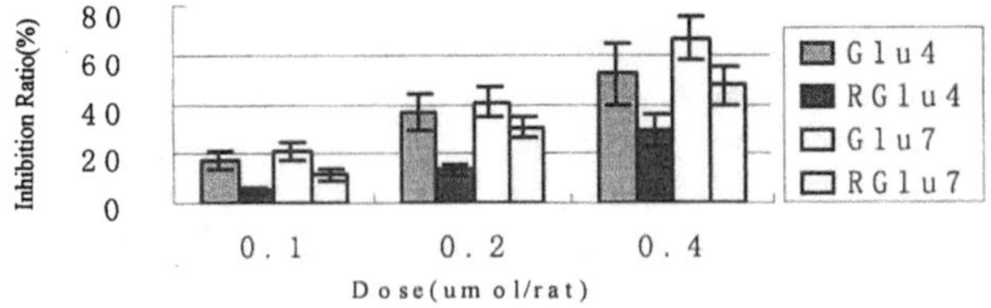

Fig. 2. The effect of deltorphin analogues on Ang II induced hypertension.

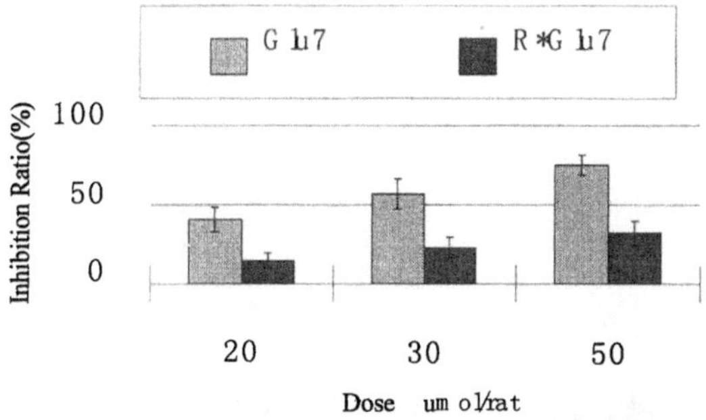

Fig. 3. The effect of deltorphin analogues on NE induced hypertension.

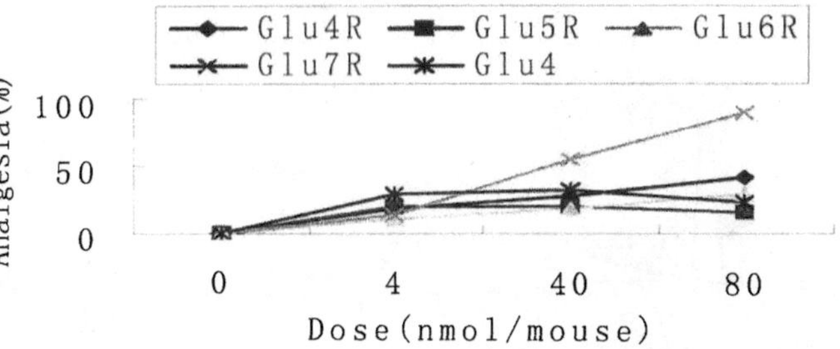

Fig. 4. Dose analgesia effect lines of the spin-labeling derivatives.

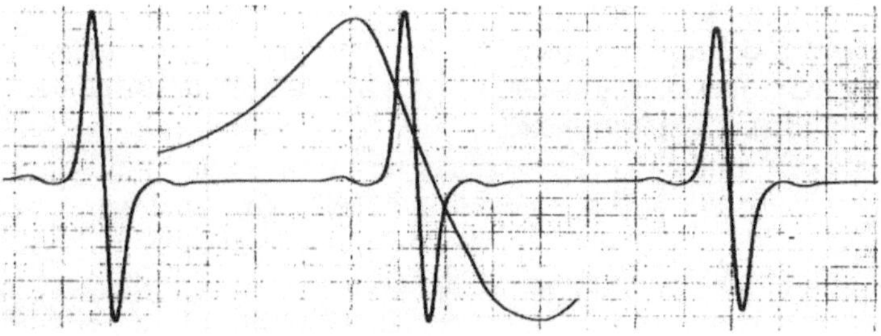

Fig. 5.1 ESR spectrum of R.

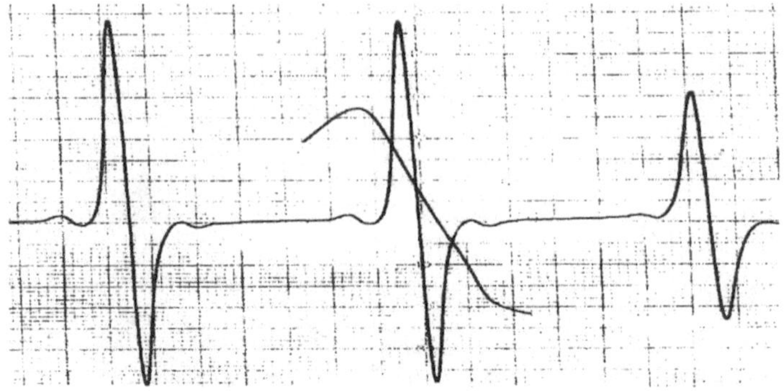

Fig. 5.2 ESR spectrum of R[Glu4]Del II.*

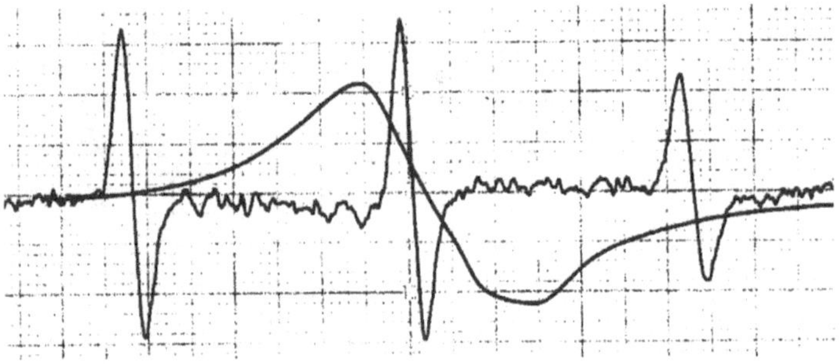

Fig. 5.3 ESR spectrum of R[Glu4]Del II*Receptor.*

The effects of deltorphin analogues on Ang II and NE induced hypertension indicated that the effect of spin-labeling on the peptide follows no set rule. From the ESR spectrum of the labeling material (fig. 5.1) we found there were three almost identical high peaks, which is characteristic of nitroxigen free radical. The 3rd peak of spin-labeling peptide (fig. 5.2) was apparently decreased, which means that the labeling was successful. In Figure5.3 the 3rd and the first peak has change than Figure 5.2,the distance between the peaks differed but not critically, so we infer that the binding site was wide and the N-terminal was on the outside. Those results agree with Deborall L H et al [5].

In summary, spin-labeling can affect some bioactivities of parent peptides, but the effects are randem. ESR technique is a useful method in studying the interaction of peptides and their receptors.

Acknowledgements

This work was supported by grants from the National Natural Science Foundation of China (No: 29442010, No: 29502007), Fok Ying Tung Education Foundation and the State Education Commission of China.

References

1. Ohnishi, S. and McConnel, C.F., J. Am. Chem. Soc., 87 (1965) 2293.
2. Stone, T., Proc. Nati. Acad. Sci., 54 (1965) 1010.
3. Hasseth, R.C., J. Med. Chem., 37 (1994) 1572.
4. Vittori, E., Proc. Nati. Acad. Sci., 86 (1989) 5188.
5. Deborall, L.H., Int. J. Peptide. Protein. Res., (1992) 450.

Design and synthesis of salmon calcitonin and its analogues

He-Ping Pan, Liang-You Wang, Zheng-Ying Chen, Fang Wang and Hui-Xin Wang

Institute of Basic Medical Sciences, Beijing, 100850, China

Introduction

Calcitonin is a peptide hormone of 32 amino acid residues with a disulfide bridge between positions 1 and 7. Salmon calcitonin (sCT) exhibits the most potent hypocalcemic effect of calcitonin species, such as human calcitonin (hCT), eel calcitonin (eCT) and pork calcitonin (pCT). Studies have demonstrated: 1) the N-terminal disulfide bridge is not required for the biological activity of sCT [1], 2) after deletion of some residues of sCT the essential biological effect is retained [2], 3) substitution of Val^{8} with Gly^{8} can eliminate the centrally and peripherally anorectic action of sCT [3].

Results and Discussion

To confirm and further study the structure and bioactivity relationship of salmon calcitonin, we designed and synthesized sCT(I) and its six analogs including [Cys $(Acm)^{1,7}$] sCT (II), [Cys $(Acm)^{1}$, Ala^{7}] sCT (III), [Gly^{8}] sCT (IV), [$Ala^{1,7}$, Gly^{8}] CT (V), [Cys $(Acm)^{1}$, Ala^{7}, Gly^{8}, des (Leu^{19}, Gln^{20})] sCT (VI), [Gly^{8}, des (Leu^{19}, Gln^{20})] sCT (VII). All the peptides were synthesized by N^{α}-Fmoc-amino acid standard solid phase procedure using HBTU-HOBT-NMM as coupling reagent. The peptides were cleaved from the resin with TMBS/Thioanisole/TFA in the presence of two additional scavengers, m-cresol and EDT. The disulfide bridges of sCT (I), sCT (IV), and sCT (VII) were formed by potassium ferricyanide oxidation. The peptides were purified by RP-HPLC. The structure of all target peptides were confirmed by amino acid analysis. The primary pharmacological observations indicates that all synthetic peptides have obvious hypocalcemic activities in intact rats.

Table 1. Amino acid composition analysis results of synthetic sCT and its analogs.

	Asp	Thr	Ser	Glu	Pro	Gly	Ala	Val	Leu	Tyr	His	Lys	Arg
SCT(I)	1.9(2)	4.6(5)	3.4(4)	3.3(3)	1.9(2)	3.0(3)	—	1.2(1)	4.6(5)	1.0(1)	1.0(1)	2.0(2)	1.0(1)
II	1.8(2)	4.8(5)	3.5(4)	3.3(3)	2.1(2)	2.8(3)	—	1.0(1)	4.9(5)	1.0(1)	1.0(1)	2.0(2)	1.0(1)
III	1.8(2)	5.1(5)	3.7(4)	3.4(3)	2.1(2)	2.9(3)	0.9(1)	1.1(1)	5.2(5)	1.0(1)	1.1(1)	2.2(2)	1.0(1)
IV	1.8(2)	4.7(5)	3.4(4)	3.4(3)	2.7(2)	3.8(4)	—	—	4.8(5)	1.0(1)	1.0(1)	2.1(2)	1.0(1)
V	1.7(2)	4.9(5)	3.6(4)	3.3(3)	2.1(2)	2.9(3)	1.8(2)	—	4.8(5)	1.1(1)	1.0(1)	2.0(2)	1.0(1)
VI	1.7(2)	4.8(5)	3.5(4)	2.2(2)	2.0(2)	3.6(4)	0.9(1)	—	3.8(4)	1.0(1)	1.0(1)	1.9(2)	1.0(1)
VII	1.6(2)	4.6(5)	3.3(4)	2.1(2)	2.7(2)	3.7(4)	—	—	3.7(4)	1.0(1)	1.0(1)	2.0(2)	1.0(1)

Table 2. Hypocalcemic effect of synthetic sCT and its analogs in rats.

Peptides	Hypocalcemic effect (iu/mg)
SCT (I)	5000
$[Cys(Acm)^{1,7}]sCT$ (II)	4010
$[Cys(Acm)^{1},Ala^{7}]sCT$(III)	3380
$[Gly^{8}]sCT$ (IV)	4189
$[Ala^{1,7},Gly^{8}]sCT$ (V)	3872
$[Cys(Acm)^{1},Ala^{7},Gly^{8},des(Leu^{19},Gln^{20})]sCT$ (VI)	1456
$[Gly^{8},des(Leu^{19},Gln^{20})]sCT$ (VII)	1779

References

1. Orowski, R.C., Epand, R.M. and Stafford, A.R., Eur. J. Biochem., 162 (1987) 339.
2. Epand, M.R., Epand, R.F., Stafford, A.R. and Orlowski, R.C., J. Med. Chem., 31 (1988) 1595.
3. Nakamuta, H., Koida, M., Ogawa, Y. and Orlowski, R. C., Nippon Yakurigaka Zasshi, 89 (1987) 191.

Studies on the synthesis and antithrombosis activity of two peptides derived from the molecular-binding site of fibrinogen

Xiao-Yu Hu [a], Ke-Qin Li [b], Guo-Ling Yang [a] and Chun-Hai Li [b]
[a] *School of Life Science, Lanzhou University, Lanzhou, 73000;* [b] *Department of Tumor Molecular Biology, the Institute of Basic Medical Science and Taiping Hospital, Beijing, 100850, China*

Introduction

Two sites on fibrinogen play an important role in thrombosis. The first site is the RGD sequence located in the Aα95-97.Aα572-574 area of fibrinogen, which binds to the platelet fibrinogen receptor gpIIb/IIIa [1]. The second site is the GPR sequence which has been found at the terminal amino of the fibrin α chain (Aα 17-19 of fibrinogen). This sequence appears to be the minimum structure which binds to the receptors on the fibrinogen polymerization site [2]. Based on these observations, we synthesized two peptides derived from the sites, RGDS and HPA (GPRVVE Hexylpeptide modified by Anisole). Antithrombosis activity was studied in *vitro* and *vivo* and although both had antithrombotic activity, the mechanism of their biology activity differed.

Methods and Results

Peptide synthesis

RGDS synthesis: NH_2-Arg-Gly-Asp-Ser-Resin was synthesized by manual solid phase peptide synthesis on Boc-Ser modified Merrifield resin (s=0.50 mmol/g Resin) with Boc strategy in the order of peptide synthesis. The peptide was cleaved from the resin using anhydrous HF [1/10(v/v)=Anisole/HF, reaction temperature is 0~4 °C], peptide was purified with Sephadex G-15 column. Analyical HPLC indicated the purity of the peptide is 90.7%. The molecular weight was determined by FAB-MS and was consistent with that calculated [$(M+1)^+$=434M/Z]. Amino acid analysis demonstrated the peptide was: Arg(1.01), Gly(0.96), ASP (1.01), Ser(1.00). The product is: NH_2-Arg-Gly-Asp-Ser-COOH.

HPA synthesis: NH_2-Gly-Pro-Arg-Val-Val-Glu-Resin was synthesized by SPPS on BHA resin (s=0.45 mmol/g Resin) with Boc strategy in the order of peptide synthesis. The Hexylpeptide was cleaved from the resin by anhydrous HF and modified by Anisole at the same time [1/1(v/v)=Anisole/HF, reaction temperature is 30 °C]. The crude peptide were purified with Sephadex G-15 column and analytic HPLC indicated it purity as 85.5%. The ultraviolet absorption spectrum of the peptide demonstrated the existence of the modification group [peak one: λ=272.8 nm ξ=0.533, peak two: =218.4 nm ξ=0.937]. The molecular weight and amino acid analyses were in agreement with the calculated values [$(M+1)^+$=746M/Z; Gly(1.00), Pro(0.81), Arg(0.89), Val(1.88), Glu(1.01)]. The product is NH_2-Gly-Pro-Arg-Val-Val-Glu(Anisole)- $CONH_2$.

Antithrombosis activity assay

The effects on experimental thrombosis: Healthy rabbits (n=6 weight 2.5 ± 0.5KG) were anesthetized with pentobarbital sodium (25 mg/KG), the right neck main artery and left neck vein were isolated and intubated by a polyethylene pipe containing a 6cm length of silk string . Each rabbit was intravenously injected with peptides (8.0 µmol/Kg or 2.5 µmol/Kg HPA or 4.5% sodium chloride as control) within 10 minutes. The blood circulation was opened, circulated from the artery through the polyethylene pipe to the vein; 15 minutes later, the blood was cut off and the thrombus and silk string were removed from the pipe. The thrombus was weighed and the IR (inhibit rate) calculated as: *iv* RGDS, IR=51.82%, *iv* HPA, IR =31.97%.

The effects on fibrinolysis: Before injecting the peptides, some blood was drawn from the rabbits (n=6) as control sample. Peptides (HPA 8.0 µmol/Kg, or RGDS 8.0 µmol/Kg) were injected intravenously into the rabbits; 15 minutes later blood samples were draw from the same animals. The "Congo Red chromatic method"[3] was used to measure the fibrinolytic the blood sample. Compared with the control (23.3±1.0 U), the enzyme unit of the *iv* HPA peptide samples was 31.9 ± 1.9 U, a significant increase (ΔU= 8.6±1.4, $p<0.01$), but that of *iv* RGDS peptide was unchanged (23.0 ± 1.5 U).

The effects on blood platelet aggregation induced by ADP [4].

Human blood was collected into test tubes containing 1/10 (v/v) 3.8% trisodium citrate. Platelet-rich plasma (PRP) was obtained by centrifugation of this mixture at 800 rpm for 10 minutes and the platelet count was adjusted to 150-200 × 10^9 /L. The various concentrated peptides (HPA or RGDS) were incubated with PRP for 5 minutes at 37 °C before addition of ADP (1 *u*mol/ml). The aggregation studies were performed in a "Blood Auto-Aggregometer" within 5 minutes. The inhibition effects were calculated by the IC_{50} (50% inhibit concentration). The IC_{50} of RGDS was 0.25 mM, while that of HPA was 0.20 mM.

Discussion

In our work, two peptides derived from the molecular-binding site of fibrinogen were synthesized. One was RGDS, the other was HPA (GPRVVE Hexylpeptide modified by Anisole). HPA was obtained by a special reaction, in which cleavage and modification took place at the same time. We thought that this reaction method was applicable to the Glutamic acid-rich synthetic peptides.

The results indicated that the antithrombosis mechanism of RGDS only inhibited platelet aggregation, but the antithrombosis mechanism of HPA not only prevented platelet aggregation but also promoted fibrinolysis. Although the two peptides derived from the same fibrinogen molecule [5], they are structurally dissimilar, and were dissimilar in their antithrombosis mechanisms.

In this paper, the results indicating that RGDS inhibits platelet aggregation and that HPA promotes fibrinolysis were consistent with other works [2, 6]. However, the result that HPA has the ability to prevent blood platelet aggregation was reported by us for the first time, suggesting that HPA could interfere directly with fibrinogen binding to the platelet receptor molecule and may also exert an effect on receptor induction.

References

1. Plerschbacher, M.D., Nature, 30 (1984) 309.
2. Bakdhy, A. K., Int. J. Peptide Protein Res., 35 (1990) 73.
3. Zhu, L.H., J. of Chinese Medical Test, 12 (1989) 42.
4. Krishnamurthi, S., Biochem. Biophys. Res. Commun., 163 (1989) 1256.
5. Edward, F., Biochemical Pharmacology, 36 (1987) 4035.
6. Cheryl, A., Int. J. Peptide protein Res., 45 (1995) 290.

Synthesis and expression of ω-conotoxin MVIIA

Min-Ying Cai and Xian-Ming Qu
Shanghai Research Center of Biotechnology Chinese Academy of Sciences, No.500, CaoBao Rd, Shanghai, 200233, China

ω-conotoxin(ω-CTX) is a peptide isolated from marine cone snails. This toxin is a blocker of presynaptic calcium channels in the central nervous system. The size of these toxins ranges from 24-29 residues. They are amidated, positively charged and have three disulfide bridges. The specific mode of action of the ω-CTX is to block presynaptic voltage-sensitive Ca^{2+} channels, and thus to inhibit transmitter release [1]. Up to now, ω-conotoxin MVIIA has been successfully used in aerosols, sachets, or lozenges by patients with spasmodic asthma, tracheitis, or bronchitis [2]. This paper focus on the study of ω-CTX MVIIA synthesis and expression in *E.coli* with the vector of pT_7ZZa.

Introduction

Biochemists have long recognized that the potency of many toxins is related to their highly specific interference with crucial parts of the machinery underlying cellular processes. Toxins can thus be used as tools for probing biochemical pathways and mechanisms. Conotoxin is produced by marine snails of the genus *Conus*. The snails comprise about 300 species of venomous predators found in tropical waters, many of them prized for their beautiful shells. The *Conus* members produce a set of small, disulfide-rich, calcium channel-targeted peptides, features common to all □-conotoxins. The characteristic arrangement of cysteine residues-C....C......CC...C......C-is called the "four-loop Cys scaffold" of the conotoxins[3]. The extent of structural and functional diversity in these groups of molecules, and their potential as biochemical tools in neurophysiology has become apparent. Derivation of polypeptides from protein precusors and their direct expression from the short synthetic genes may be unsuccesssful for different reasons. To study expression of the small genes we have synthesized the DNA sequences of ω-CTX MVIIA and constructed a recombinant expression plasmid encoding a protein A ω-CTX MVIIA fusion protein. The fused ω-CTX MVIIA expressed in the *Escherichia coli* and can be purified with the IgG Sepharose 6 Fast Flow.

Materials and methods

Enzymes and chemicals kits

Restriction endonucleases, DNA polymerase and DNA ligase were purchased from Promega; Wizard™ *Plus* Miniprips DNA Purification System was from Promega

(Madison, WI, U.S.A.), Gel Extraction Mini kit was from Watson Biomedical, Inc; X-gal and IPTG were from Sigma; DNA marker were from Sino-American Biotechnology Company.

IgG Sepharose 6 Fast Flow came from Pharmacia Biotech; factor Xa was from Promega; Protein marker from Shanghai DongFeng Factory, lower molecular protein marker was from MERCK.

Bacterial strains and plasmids

Escherichia coli strains: BL21: F^-, ompT, $hsdS_B$,(r_B^-, m_B^-)dcm, gal ;BL21(DE3) ompT, $hsdS_B$, (r_B^-, m_B^-) dcm, galλ (DE3); DH5α: φ80dlacZΔM15, recA1, endA!, gyrA96, thi-, hsdR17 (r_k^-,m_k^+), supE44, relA1, deoR, □(lacZYA-argF)U169, all of these strains were keep in our lab.

plasmids: pUC18 were keep in my lab, pT_7ZZa was kindly provided by Stefan Stahl's lab (Royal Institute of Technology, Sweden).

Construction of recombinant plasmid pT7ZZ ω-ConotoxinMVIIA

Synthesis of the oligonucleotides: Codon usage table [4-5] adapted both for E.coli and yeast were chosen as the oligonucleotides. The synthesis of four oligonucleotides (fig. 1) were processed by the Sangong Inc.

1. 5'AAT TCA TCG AAG GTC GTT GTA AAG GTA AGG GTG CTA AAT GTT3'
2. 5'CTC GTC TGA TGT ACG ATT GTT GTA CTG GTT CTT GTC GTT CTG GTA AGT GTT AAT GA3'
3. 5' AGC TTC ATT AAC ACT TAC CAG AAC GAC AAG AAC CAG TAC AAC AAT CGT ACA T3'
4. 5'CAG ACG AGA ACA TTT AGC ACC CTT ACC TTT ACA ACG ACC TTC GAT G3'

Fig. 1. The sequences of four oligonucleotides.

The gene sequence was deduced from a published paper [6]. There is TAA, TGA stop codon at the 3'end of the ω-conotoxin MVIIA; at the N-terminal of the ω-CTX we added the sequence of Factor Xa and their 5'-ends contain sticky ends for the EcorRI restrictase and the Hind III enzyme at 3'-ends. Phosphorylated oligonucleotides were annealed and ligated into pUC18 plasmid treated with HindIII and EcoRI enzymes. The gene sequence was identified by the Shanghai HaoJia International Trading Co.Ltd. Fragment contained the ω-CTX MVIIA and HindIII, EcoRI was ligated into pT7ZZa plasmid also treated with HindIII and EcoRI enzymes (fig. 2). Recombinant plasmid pT7ZZ ω-CTX MVIIA was analyzed by HindIII, EcoRI and Rsa I enzymes. It was proved that recombinant plasmid was positive to the Rsa I treatment as analyzed by 2.5% Agarose gel electrophoresis (fig. 3). All genetic procedures were carried out according to the manufacturer and published manual (Sambrooke et al., 1989)

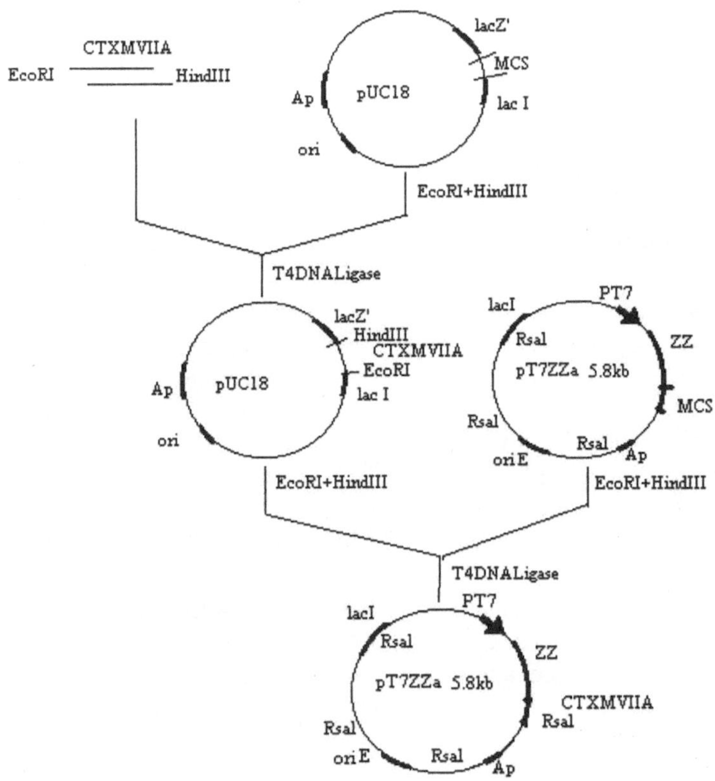

Fig. 2. Construction of the plasmid of pT7ZZ ω-conotoxin (ω-CTX).

Expression of ω-conotoxin MVIIA

For the expression of ω-CTX MVIIA in *E.coli* we used a previously described protocol [7]. Overnight culture of BL21 (DE3) cells transformed with pT7ZZa were inoculated 1: 100 into 250 ml LB medium containing 50 μg/ml ampicillin. Cultures were grown to an A_{600} of 0.6-0.8 with continuous shaking at 37 °C. Transcription of the cloned sequence was induced with IPTG added to a final concentration of 0.1mM. Following a 3 h induction, cells were harvested by centrifugation, resuspended in sonication buffer (300 mM NaCL/50 mM sodium phosphate, pH 8), sonicated 2min and centrifuged at 28000 g for 30 min at –4 °C, and the supernant was collected. The pellet was resuspended in sonication buffer. PMSF was added to a final concentration of 1mM in the pellet and supernant to prevent anti-degradation of protein. The expression of ω-conotoxin MVIIA was analyzed by 15% SDS-PAGE [8-9].

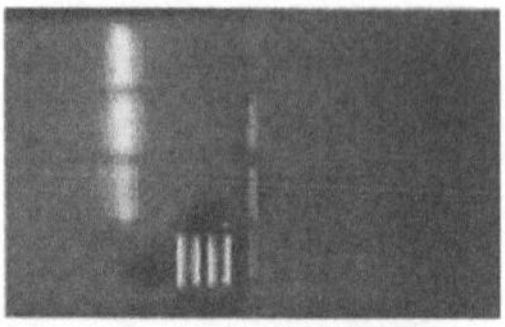

Fig. 3a. Recombinant plasmid pUC18 ω-CTX MVIIA was treated by the EcorI and Hind III.

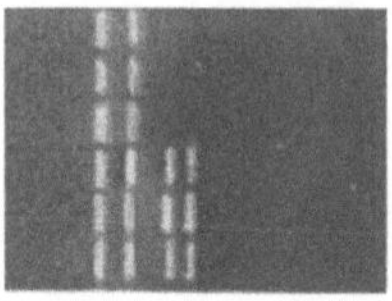

Fig. 3b. Recombinant plasmid pT7ZZ ω-conotoxin MVIIA were treated with Rsa I (four fragments).

Purification of recombinant ω-CTX MVIIA and Removal of the protein A

Recombinant ω-CTX MVIIA containing a ZZa region at the N-terminus was affinity purified on IgG Sepharose 6 Fast Flow (Pharmacia Biotech) and washed out as recommended by the manufacturer. The fusion protein was collected and dried at –70 °C.

The fusion protein contained at the cleavage site of Factor Xa at the N-terminal of ω-CTX MVIIA. We used 2% Factor Xa at 22 °C for 6hr to cleave protein A. G-50 Sephardex was chosen to isolate the ω-ConotoxinMVIIA from Protein A.

Results and discussion

Cloning and expression of recombinant ω-Conotoxin MVIIA

The size of plasmid pT7ZZa having T7 promoter is 5.8 k. Downstream of ZZ is the recognition site of HA and His64Ala-subtilisin, enabling release of target proteins produced as ZZ-fusions, recovered by affinity chromatography on IgG-Sepharose. The expression strain is BL21 (DE3). After induction for 3hr with IPTG, the expression of fusion protein was the highest. Longer induction time resulted in lost plasmid. Therefore, the expression yield controlled by the T7 promoter is time limited. Fig. 4a shows expression of recombinant plasmid transformed BL21 (DE3) induced by 0.1mM IPTG. After 3hr induction the expression yield of fusion is highest enhanced by Laser Densitometer (LKB). The pellet and supernant were analyzed by SDS/15%-PAGE gel; up 90% of the fusion protein is dissolvable in the cell. There is no degradation. So we deduce that the fusion protein is mainly expressed *in vivo*. Perhaps the strain of BL21 (DE3) ompT, $hsdS_B$, (r_B^-, m_B^-) is deficient in protein enzyme and the fusion protein is not toxic to the strain.

Isolation of fusion proteins and Factor Xa cleavage

After cleaving Protein A with Factor Xa, we obtained the free ω-conotoxin MVIIA. The representative Coomassie Brilliant Blue R250-stained SDS/15%-PAGE gel shows expressed ω-conotoxin MVIIA extracted from a 250 ml culture (fig. 4a). SDS/13.8%-PAGE (8M Urea) gel shows purified ω-conotoxin MVIIA Amide and Patch-clamp technology to detect the activity of ω-CTX MVIIA is in preparation.

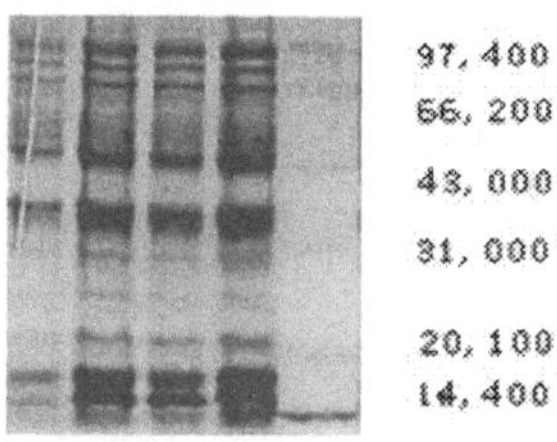

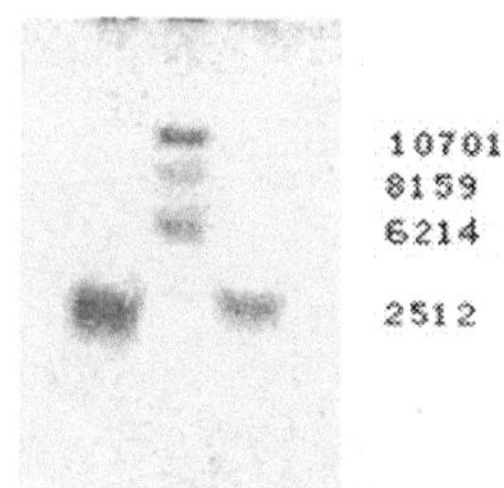

Fig.4a Coomassie Brilliant BlueR250-stained SDS/15%-PAGE gel of different times expressed fusion ω-CTX MVIIA, Each lane was loaded with 20 μl of crude lysate or protein fraction. Lane 1:before induction, Lane 2: induced 1hr, Lane3: induced 2hr, Lane 4: induced 3hr, Lane 5: Protein Marker.

Fig.4b Coomassie Brilliant Blue R250-stained SDS/13.8%-PAGE (8M urea) of ω-CTX MVIIA.

Considering all the above data we can conclude that a short peptide such as□-CTX MVIIA (with certain sequences) not only can be chemically synthesized but also can be biosynthetically developed in prokaryotic systems, which may result in an economical pathway to synthesis of small peptides (20-40 residue).

References

1. William, R.G. and Baldomero, M.O., Ann. Rev. Biochem., 57 (1988) 665.
2. GB (1993) 2262886.
3. Baldomero, M.O., Ann. Rev. Biochem., 63 (1994) 823.
4. Shiping, Z.G. and Zubay, E. G., Gene, 105 (1994) 61.
5. Waada, K.N., Wada, Y., Ishibashi, F. and Gojobon, T., Nucleic acids Res., 20 (1992) 2111.
6. Baldomero M., Olivera, W.R., Gray, R., Zeikus, J.M. and Mcintosh, J., Science, 230 (1985) 1338.
7. Li, B. -Y., Frankel, A.E. and Ramakrishnan, S., Protein Expression Purif., 3 (1992) 386.
8. Wahlsten, J.L. and Ramakrishnan, S., Biotechnol. Appl. Biochem., 24 (1996) 155.
9. Martemyanov, K.A., Spirin, A. S. and Gudkov, A.T., Biotechnol. Lett., 18 (1996) 1357.

Session D
Peptide vaccines/Immunology/Virus

Chairs: Xiao-Yu Hu
Department of Biochemistry and Molecular Biology
School of Life Science
Lanzhou University
Lanzhou, China

B. Kutscher
ASTA Medica AG
Dresden, Germany

Jie-Cheng Xu
Shanghai Institute of Organic Chemistry
Chinese Academy of Sciences
Shanghai, China

Studies on the synthetic peptide vaccines against schistosomiasis

Sheng-Li Cao[a], Zhi-Hui Qin[a], Meng-Shen Cai[a] and You-En Shi[b]
[a]*School of pharmaceutical sciences, Beijing medical University, Beijing, 100083;*
[b]*School of basic medical sciences, Tongji medical university, Wuhan, 430030, China*

Introduction

Developing an effective synthetic peptide vaccine is an important strategy for the immunological control of schistosomiasis, one of the major parasitic diseases seriously affecting human health, especially in developing countries. On the basis of the amino acid sequences of the schistosomal host-protective antigens Sm28GST and Sj26GST, four antigenic peptides, three of which (P_{26}, P_{116} and P_{141}) derived from Sm28GST and one (J_{187}) derived from Sj26GST, were previously synthesized and screened by ELISA in our laboratory [1, 2]. In most cases, small synthetic peptides must be conjugated to a carrier protein to render them immunogenic. However, since protein carriers may cause carrier-induced toxicity and epitopic suppression, they are undesirable for human vaccines.

In order to avoid the use of a carrier molecule and to increase immunogenicity, we designed five MAP compounds containing the same peptides using the concept of multiple antigen peptide initially described by Tam [3, 4]. Moreover, it is now clear that resistance to schistosome infection requires both humeral and cellular immunity. Therefore, an ideal synthetic peptide vaccine against schistosomiasis should be highly immunogenic and able to induce both T and B cell schistosome-specific immune responses [5]. For these reasons, we designed six additional MAPs composed of two different peptides.

Results and discussion

The MAP is a macromolecular compound with unambiguous structure consisting of an oligomeric branching lysine core and dendritic peptide arms. Boc chemistry was used to synthesize the 5 MAPs composed of the same peptides, whereas both Boc and Fmoc chemistry were used to synthesize the 6 MAPs composed of two different peptides. The MAP-resin was cleaved with low-high HF method to remove the MAP from the resin support and the crude peptide was purified batch-wise by gel-permeation chromatography. Amino acid analyses gave results in agreement with the expected composition.

The antigenicities of the synthetic MAPs were examined with dot-ELISA and the results indicated those the immunoreactivities of MAPs with specific antibodies were higher than these of the corresponding monomeric peptides. In the absence of any immunological adjuvant, all the synthetic MAPs were able to elicit antibody responses and were able to react with the mixture of soluble egg antigens and adult worm antigens in both KunMing and BALB/c mice. Furthermore, mice vaccinated with some of the synthetic MAPs were significantly protected against a challenge infection with *Schistosoma japonicum* cercariae. The experiment results are listed in tables 1 and 2.

Table 1. Protective efficacy in KunMing Mice Vaccinated with the synthetic MAPs.

Peptide	Animal number	Mean worm burden±SD	WRR* (%)	Mean egg number/1g liver (±SD)	ERR** (%)
P_{26}-MAP_4***	10	13.4±2.7	59.9	14718±4906	61.1
P_{141}-MAP_4	10	11.9±2.0	64.4	17083±4587	54.9
J_{187}-MAP_4	7	19.6±4.6	41.3	20603±8393	45.6
$P_{26}P_{141}$-MAP_4	8	13.0±5.6	61.1	12452±2508	67.1
$P_{26}P_{141}$-MAP_8	8	21.9±3.8	34.4	19921±11187	47.4
$P_{26}J_{187}$-MAP_4	9	20.0±8.8	40.1	19536±4609	48.4
Control	10	33.4±2.1		37884±7546	

** WRR: worm reduction rate ** ERR: egg reduction rate *** The subscript of MAP represents the number of peptides in the MAP molecule.*

It can be seen that immunization with the synthetic MAPs achieve two goals in our experimental animals: a) a significant reduction in worm burden resulting from infection, and b) a significant decrease in liver tissue eggs, which would reduce pathological consequences. Since schistosomes do not replicate within the tissue of their definitive hosts, a vaccine that is effective in reducing worm burden by 50% or more can considerably reduce pathology and affect parasite transmission [6]. Furthermore, it is the viable eggs that cause disease pathology. Therefore, some of these MAPs will be useful for the development of effective anti-schistosome vaccines.

Table 2. Protective efficacy in BALB/C Mice Vaccinated with the synthetic MAPs.

Peptide	Animal number	Mean worm number±SD	WRR (%)	Mean egg number/1g liver ±SD	ERR (%)
P_{26}-MAP_4	9	18.0±4.6	37.5	35067±13710	35.1
P_{116}-MAP_4	8	10.8±4.6	62.5	24850±6881	54.0
P_{141}-MAP_4	9	25.4±3.1	11.8	40300±10118	25.4
P_{26}-MAP_8	8	20.7±3.2	28.1	32100±7803	40.6
$P_{26}P_{141}$-MAP_4	10	19.4±3.5	32.6	33150±7708	38.7
$P_{26}J_{187}$-MAP_8	8	15.0±2.1	47.9	26787±6291	50.4
$P_{26}P_{116}$-MAP_8	9	7.6±4.3	73.6	13022±7442	75.9
$P_{116}P_{141}$-MAP_8	10	9.1±3.2	68.4	17967±7304	66.7
Control	9	28.8±4.9		54089±14026	

Acknowledgments

The project was supported by the Foundation of the Ministry of Public Health, China.

References

1. Mao, F., Doctoral dissertation, Beijing Medical University, 1991
2. Xu, J.X. and Cai, M.S.,In Lu, G,S; Du,Y.C.and Tam,T.J.(Eds.) Peptide:Biology and Chemistry (Proceeding of the 1994 Chinese peptide Symposium),ESCOM,Leiden, 1995, p.256.
3. Tam, J.P., Proc. Natl. Acad. Sci. USA, 85 (1988) 5409.
4. Posnett, D.N., McGrath, H. and Tam, J. P., J. Biol. Chem., 263 (1988) 1719.
5. Wolowczuk, I., Auriault, C. and Gras-Masse, H., J. Immunol., 142 (1989) 1342.
6. Dunne, D.W., Hagan, P. and Abath, F.G.C., Lancet, 345 (1995) 1488.

Studies on the anti-schistosomal synthetic peptide vaccine: Prediction and synthesis of antigenic peptides of *Schistosoma* paramyosin

Jia-Xi Xu* and Jun-Hai Yang
Department of Chemistry, Peking University, Beijing, 100871, China

Introduction

Schistosomiasis is a debilitating parasitic disease affecting more than 200 million people throughout the world and is responsible for 800,000 deaths per year according to recent estimations of the World Health Organization [1]. The disease, which causes major liver damage in humans and is potentially fatal if untreated, results from infection by one of three major schistosome species, *Schistosoma mansoni, Schistosoma haematobium* and *Schistosoma japonicum*. The three parasites differ in a number of characteristics including morphology and pathogenicity, as well as in their geographical distribution. *Schistosoma mansoni* is the most widespread, occurring in Africa, the Middle East and South America. *Schistosoma haematobium* is present in African and some Middle-Eastern countries, while *Schistosoma japonicum* is found in Asia, particularly China and the Philippines [2]. It is necessary to study an effective vaccine inducing significant levels of protection against the invasive stage of the parasite. It is very important to seek effective antigenic peptides from various protective antigens of schistosomes. Paramyosin, a 97kDa polypeptide, is an effective antigen candidate with 60% to 77% protection in mouse, the highest protection among known candidate antigens [3-6]. The antigenic peptides of 97kDa paramyosins of *Schistosoma mansoni* and *Schistosoma japonicum* have been predicted and synthesized and their antigenicities have been determined by Dot-ELISA method.

Results and Discussion

Prediction of epitope
Nucleotide sequences and amino acid sequences of the 97-kDa paramyosins of *Schistosoma mansoni* and *Schistosoma japonicum* were obtained from published data [2]. Their epitopes were predicted according to their hydrophilicity (Hopp and Woods method), flexibility, and accessibility by a computer program PC-Gene [7-9]. The secondary structures of two paramyosins were predicted by the Chou and Fasman method [10]. The sequences of the epitopes are listed below.

Sm97-1 SLDDEAKN
Sm97-2 DEESEEA
Sm97-3 EEFEEMKRK
Sm97-4 LTSQDSQL

Sm97-5 RLRERDEELE
Sm97-6 LRKSTTRTI
Sj97-1 SLDDESRN
Sj97-4 TSQNNQL

The sequences of Sj97-2, Sj97-3, Sj97-5 and Sj97-6 are the same as those of Sm97-2, Sm97-3, Sm97-5 and Sm97-6, respectively.

Synthesis of epitopic peptides

These predicted peptides have been synthesized using the Merrifield solid phase peptide synthesis method with the acid-labile tert-butyloxycarbonyl (Boc) group for temporary protection and acid-stable groups for the protection of side chains [8, 9, 11]. Side chain-protected peptide-resins were cleaved by anhydrous hydrogen fluoride under anisole, 1,2-ethandithiol and thioanisole as scavengers and washed with cooled ethyl ether. Peptides were extracted with 30% acetic acid and purified by gel filtration on Sephadex G15. They were checked for homogeneity by reverse-phase HPLC. Impure peptides were further purified by preparative HPLC. The compositions of the peptides were confirmed by amino acid analysis and FAB-MS.

The synthetic antigenic peptides Sm97-3 and Sj97-4 showed antigenicities in dot blot immunobinding assay [12].

Acknowledgments

Project 29772002 supported by National Natural Science Foundation of China.

References

1. Sher, A., James, S. L., Correa-Oliveira, R., Hieny, S. and Pearce, E., Parasitol., 98 (1989) s61.
2. Yang, W., Waine, G. J., Sculley, D. G., Liu, X. and McManus, D. P., Int. J. Parasitol., 22 (1992) 1187.
3. Xu, J. X., Chemistry of Life, 16 (1996) 42.
4. Flanigan, T. P., King, C. H., Lett, R. R., Nanduri, J.and Mahmoud, A. A. F., J. Clin. Invest., 83 (1989) 1010.
5. Lanar, D. E., Pearce, E. J., James, S. L. and Sher, A., Science, 234 (1986) 593.
6. Pearce, E. J., James, S. L., Hieny, S., Lanar, D. E. and Sher, A., Proc. Natl. Acad. Sci. USA, 85 (1988) 5678.
7. Hopp, T. P. and Woods, K. R., Proc. Natl. Acad. Sci. USA, 78 (1981) 3824.
8. (Xu, J. X. and Cai, M. S., Chin. Chem. Letts., 9 (1998), unpublished.).
9. Xu, J. X., Cai, M. S. and Shi, Y. E., Chem. J. Chinese Univ., 17 (1996) 424.
10. Chou, P. Y. and Fasman, G. D., Biochemistry, 13 (1974) 222.
11. Merrifield, R. B., J. Am. Chem. Soc., 85 (1963) 2149.
12. Boctor, F. N., Stek, J., Peter Jr, J. B. and Kamai, R., J. Parasitol., 73 (1987) 589.

Construction and expression of a recombinant antibody targeted plasminogen activator

Jia-Shu Yang, Peng-Chen Jiang and Bing-Gen Ru
National Laboratory of Protein Engineering, College of Life Science, Peking University, Beijing, 100871, China

Introduction:

The occlusion of a cerebral or coronary artery by a blood clot accounts for the majority of deaths in industrialized countries and is responsible for significant incapacitation and morbidity [1]. The immediate cause of occlusion is a defect in the vessel wall, but it is the formation of a thrombus on this defect that almost invariably results in the final interruption of blood flow. Dissolution of the thrombus with plasminogen activator therapy is an attractive method of restoring blood flow.

In the past few years, plasminogen activators have been shown to be effective therapeutic agents in coronary artery thrombosis improving overall survival while limiting the degree of damage to the heart. Unfortunately, because plasmin modifies platelet function [2] and degrades circulating fibrinogen and clotting factors V and VIII [3], as well as thrombus-bound fibrin, plasminogen activator therapy also carries the risk of hemorrhage. Must the risk accompany any effect to bring about thrombolysis? Perhaps the ability to selectively rather than systemically generate plasmin would improve the safety of plasminogen activator therapy.

Antibody specific for fibrin and activated platelets could achieve an affinity and selectivity of plasminogen to the thrombus. SZ51 was a monoclonal antibody against the membrane protein, GMP140, on the activated platelets in thrombi [4]. It is thus expected to improve thrombolytic efficiency if SZ51 is linked to urokinase. In this study, we constructed a chimeric protein in which the scFv (single chain fragment of variable region) of SZ51 is linked to the scu-PA-32k (leu144-leu411) at the NH_3-terminal.

Results and Discussion

Polymerase chain reaction (PCR) was used to amplify the region of V_H and V_K from the Fab fragment of SZ51, and scu-PA-32k (leu144-leu411) from the urokinase gene, respectively. Through suitable linkers and appropriate restriction sites, these fragments were joined together in the order of V_H-V_K-scu-PA-32k and inserted into the expression vector, pGTC. The latter was a derivative of pGEX-4T-1. The restriction map of the vector and partial sequencing identified the positive clone.

Transformed with the expression vector and not induced by IPTG, the host bacteria, JM109 or DH5α, secreted the recombinant protein directly into the media along with some green material. The media became more and more dense with incubation time.

The green material and the expression product did not interfere with the bacteria's growth and the bacteria was still able to secrete the protein at a high O.D. value. Because the green material could be titrated by HCl, it might be an alkaloid. It had a small molecular weight for being dialyzed through the membrane with 13500 cut off. As the green material had no any detectable activity, it was probably a secondary metabolism product. In this system, it acted as a marker for the expression level.

The supernatant had strong fibrinolytic activity on the fibrin plate, and this activity kept the same trend with the $O.D._{396}$ of the culture (fig. 1). In ELISA, the supernatant had a positive reaction on the plate bound with the activated platelet. In Western assay, the supernatant had weaker reactions with multiclonal antibody against urokinase. In fact, the fibrinolytic activity could be inhibited by the multiclonal antibody against urokinase on the fibrin plate.

Two forms of the recombinant protein, 60kd and 38kd, appeared on SDS-PAGE with or without β-mercaptoethanol. The 38kd form had stronger fibrinolytic activity than that of the 60 kd form. During purification of the recombinant protein, the content of 38kd increased, and that of 60 kd declined. We therefore speculated that 38kd might be a degradation product of 60 kd via an unknown protease or autocatalysis, although there was previously no obvious evidence of this degradation. The two forms could be purified separately through Sephadex G-100 and Phenyl Sepharose 4 Fast Flow chromatography. The specific activity of 38kd could reach 70000IU/mg on the fibrin plate, and that of 60 kd was lower. The enhancement of fibrinolysis due to the platelet affinity of scFv is under investigation.

Some merits of the expression system can be summarized as follows:

1 No induction.

2 Did not interfere with the bacteria growth.

3 Secretion into the media is in the active form.

4 An unknown green material which can be a marker for expression is also secreted.

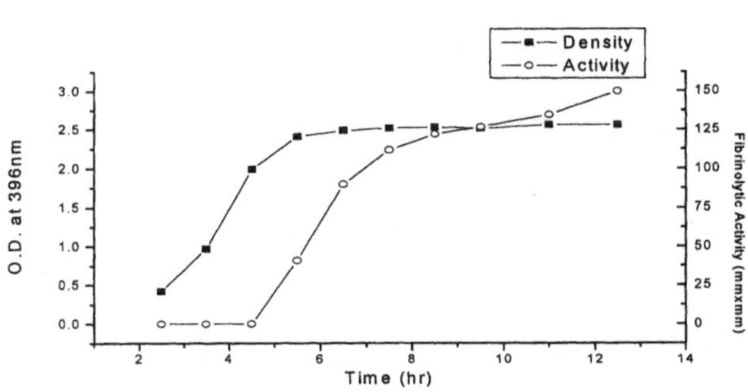

Fig. 1. The relationship between the bacteria density and the fibrinolitic activity of the supernatant.

References

1. Haber, E., Quertermous, T., Matsueda, G.R. and Runge, M.S., Science, 243 (1989) 51.
2. Stricker, R.B., Wong, D., Shiu, D.T., Reyes, P.T. and Shuman, M.A., Blood, 68 (1986) 275.
3. Marder, U.J. and Shulman, N.R., J. Biol. Chem., 244 (1969) 2120.
4. Wu, J.C., He, G.R. and Wu, G.X., NuCl. Med. Commun., 14 (1993) 1088.

Chemical conjugation between haemophilus influenzae type b (hib) polysaccharide and proteins

Ping-Sheng Lei and Gui-Shen Lu

Institute of Material Medical Chinese Academy of Medical Sciences, 1st Xian Nong Tan Street, Beijing, 100050, China

Haemophilus influenzae b polysaccharide (Hib-PS) protein conjugate vaccines differ chemically and immunologically. The structure of Hib PS was elucidated and its immunologic properties were characterized (1). Successful induction of antibody responses to *influenzae type b*-3)-β -D-Rib*f*-(1-1)- ribitol-5-PO4- polysaccharide in immunology experiment requires the conjugation of protein to the Hib-Ps. However, Hib-PS lacks chemically reactive groups which could be bound to a protein directly. Further, this polysaccharide is unstable in acidic or basic conditions.

We conjugated activated Hib-PS (carbonyl diimidazole or 6-aminohexanoic acid) with different proteins by carbodimide-mediated condensation.

The carrier proteins we used were diphtheria toxin or meningococcic vaccine. Purification of above proteins was done by extraction, centrifugation and ultrafiltration. The synthetic route is depicted in the following schemes 1 and 2.

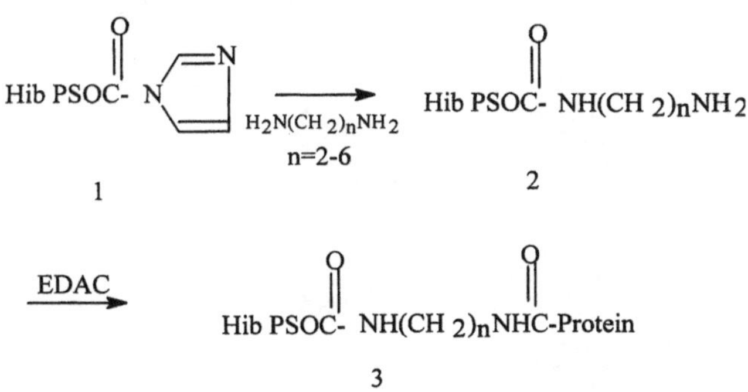

Scheme 1.

Purification of Hib-Ps protein conjugates by filtration on a column of CL-4B Sepharose and Sephadex G-25 offered target products. Hib-Ps protein conjugates with varying degrees of derivatization were prepared.

The immunological activity of Hib PS-protein conjugate was tested in mice at three doses. Groups of mice were bled 2 wks after the first immunization and the sera were assayed for Hib PS antibodies. The test showed that Hib PS-protein conjugate possessed significant immunological response after the first immunization.

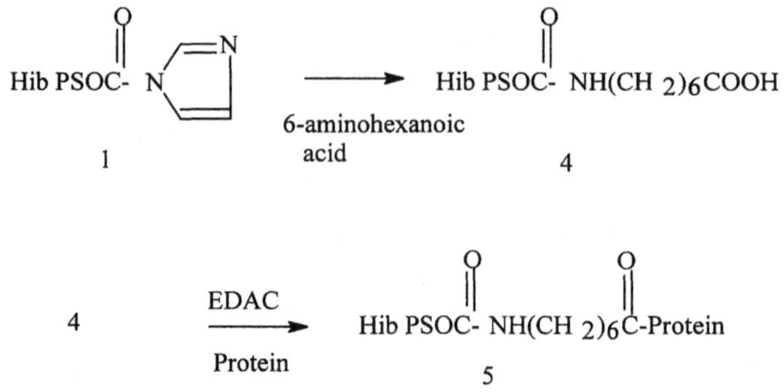

Scheme 2.

References

1. Tsui, F.P. and Egan, W., Carbohydr. Res., 173 (1988) 65.
2. Smith, D.H., Pediatrics, 52 (1973) 637.

Synthesis of human inhibin fragments, preparation and generation of monoclonal antibodies against human inhibin

Gui-Shen Lu[a], Ying Sun[a], Ping-Sheng Lei[a], Xin Huang[b], Hui-Ming Xia[b] and Xiao-Dong Gao[b]

[a]*Institute of Materia Medica, Chinese Academy of Medical Sciences, Beijing, 100050;*
[b]*Pathophisiology Department of He Nan Medical University, Zhenzhou, 450052, China*

Introduction

Inhibin secreted from gonads is a water-soluble protein consisting of a α subunit (134 residues) and two β subunits (116 115 residues), which are linked by disulfide bridges. It regulates the secretion of follicle stimulating hormone (FSH) through a feedback mechanism [1].

Serum inhibin levels change with age and reproductive cycle changes [2, 3]. There are possible sex-specific roles for inhibin A and B, and current speculation is that the major species in the male may be inhibin B. Inhibin A and B play different roles in the female [4]. Serum inhibin levels provide a diagnostic marker for determining the maturity of ovary follicle development [5], normal or abnormal pregnancy, hydatidiform mole [6] and for screening Down's syndrome [7]. Measurement of inhibin can be used as a marker for reproductive system tumors, especially for ovary granulosa cell and mucinoid cancers [7]. These findings suggest that inhibin may have diverse metabolic roles. But a complete understanding of its biology has been limited by the lack of specific and sensitive assays. The development of a specific and sensitive two-site enzyme-linked immunoadsorbent assay (ELISA) for measuring inhibin would provide greater specificity [8]. Betteridge and Craven [9], Groome, et al [4] and D. L. Baly et al [8] have reported sensitive two-site ELISA. The first report on a monoclonal antibody to inhibin prepared using synthetic peptide immunizations was by Groome et al. In order to establish a two-site ELISA, three inhibin segments: inhibin α (37-65), βA(1-28), βA(82-114) were synthesized in our laboratory. Monoclonal antibodies to separate subunits were prepared by our collaborator, Professor Hui-ming Xia.

Results and Discussion

Peptide synthesis

The fragments of inhibin subunits α(37-65), βA (1-28) and βA(82-114) were synthesized manually by stepwise solid-phase procedure on Pam resin support in Boc chemistry, and purified by HPLC. The correct products were confirmed by amino acid analysis.

Preparation of monoclonal antibodies
Different immunization methods were used for the three above-mentioned peptides.

Inhibin α (37-65)
Crude inhibin isolated from porcine follicle fluid was used as immunogen to immunize Balb/c mice. The hybridomas were cloned by growing single cell in 96-well plates. Cells in three wells were selected for cloning based on their reaction with recombinant human inhibin (32 KD) determined by ELISA. The positive hybridoma were recloned three times by limiting dilution. Three hybridoma cell lines (IA5, IV3C, V5G) were cloned which can secret monoclonal antibodies specifically to recombinant human inhibin (32 KD) and also to synthetic inhibin α (37-65). Antibody titers in ascites of three cell lines were 1:100000, 1:20000, and 1:1600.

Inhibin βA (1-28)
Inhibin βA (1-28) segment was conjugated with Tg to prepare an immune spleen cell suspension for fusion. The hybridoma culture supernants were screened by indirect ELISA with 1 μg/ml of the inhibin βA (1-28) coated segment. Those hybridomas reacting positively to the fragment were cloned and subcloned by limiting dilution over 3 times to ensure monoclonality. After 3 times fusion, we had 5 cell lines, but only one cell line (VI5C) secreting McAb reacted specifically to the inhibin βA (1-28) segment. Antibody titer in ascites was 1:4000.

Inhibin βA (82-114)
Crude inhibin isolated from porcine follicle fluid was used as immunogen to immunize Balb/c mice in the first two immunizations. In the third immunization the inhibin βA (82-114) segment was injected directly into the spleen. The hybridoma culture supernatants were screened by indirect ELISA with inhibin βA (82-114) segment as antigen. These cells reacted positively to the fragment and were cloned and subcloned by limiting dilution 3 times. After 3 times fusion, we obtained three cell lines (III7B, IV3G, II3A). Antibody titers in ascites were 1:64000, 1:16000, 1:8000.

Determination of the titer of monoclonal antibodies

Cell line	Antibody titer		
	Culture supernatant	Ascites fluid	polyantibody serum
IA5	1:6000	1:100000	1:6400
IV3C	1:1000	1:20000	1:1000
V5G	1:400	1:1600	1:1000
III7B	1:400	1:64000	1:4000
IV3G	1:600	1:16000	1:800
II3A	1:400	1:8000	1:800
VI5C	1:200	1:4000	1:400

ELISA antibody additivity test showed that McAb secreted by IA5, IV3C and V5G could recognize different epitopes. There were high amounts of immunoreactive materials in human and porcine folicular fluid measured with McAbs, but it could not be detected in the plasma of postmenopausal women. In the specificity test eight antigens which possibly cross-react with inhibin were selected. The results showed that the cross-reaction ratio was < 10%. The relative affinity was IA5 > IV3C > V5G. IA5 was used in immunohistochemistry study. The results showed that positive reactions were observed in human granulosa, theca, luteal, Sertoli, Leydig, spermatogonia and spermatocyte cells. These data were not different from other reports [10]. In tumor tissues, some gonadal tumors had inhibin immunoreactivity but some did not.

Specificity test showed that antibodies in ascites react specifically with βA subunit fragments and no cross reaction occurred with inhibin α subunit fragment, carrier protein Tg or other peptide hormones. ELISA antibody additivity test showed that McAb secreted by II7B, IV3G, and II3A recognized the same epitope in βA (82-114) fragment, and VI5C recognized the epitope in βA (1-28) fragment. The relative affinities of McAb were III7B > IV3G > II3A. II7B was used to localize the distribution of inhibin like substances. Positive reactions were observed in Sertoli, Leydig, spermatogonia and spermatocyte cells of rat testis, as well as in human ovarian granulosa cells, and granulosa tumor cells. These results were similar to other reports [11]. In tumor tissues, some gonadal tumors had inhibin immunoreactivity but some did not.

Conclusion

Several hybridoma cell lines which can secret inhibin McAb were obtained with their high specificity and sensitivity, also with their different titer. These studies were important bases for the further syudy - to establish two-site ELISA.

Acknowledgments

We are grateful to the Research Foundation of Ministry of Health for research grant.

References

1. Ying, S. Y., J. Steroid Biochem., 33 (1989) 705.
2. Mason, A. J., Niall, H. D. and Seebury, D. H., Biochem. Biophys. Res. Commun., 135 (1986) 957.
3. Robert, D. M., Foulds, L. M. and Prisk, M., Endocrinology, 130 (1992) 1680.
4. Groome, N. P., Illingworth, P. J. and O'Brien, M., Clin. Endocrinol., 40 (1994) 717.
5. Gertrud, H., Helge, N. and Carstern, S., Biochem. Biophys. Res. Commun., 206 (1995) 608.
6. Stewart, A. G., Miborrow, H. M. and Ring, J. M., FEBS Lett., 206 (1986) 329.
7. Burger, H. G., J. Clin. Endocinol. Metab., 65 (1993) 1391.
8. Baly, D. L., Allison, D. E., Krummen, L. A. and Woodruff, T. K., Endocrinology, 132

(1993) 2099.
9. Betteridge, A. and Craven, R. P., Bio. Reprod., 45 (1991) 748.
10. Khail, A., Kaufmann, R. C. and Wortsman, J., Am. J. Obstet. Gynecol., 172 (1995) 1019.
11. Vannelli, G. B., Barni, T. and Negro-vilar, A., Cell Tissues Res., 269 (1992) 221.

Session E

Peptide diversity; Chemical libraries; Ligand-receptor interaction; Peptides in therapeutics & diagnosis

Chairs: Kit Lam
Arizona Cancer Center
University of Arizona
Tucson, U.S.A.

You-Shang Zhang
Shanghai Institute of Biochemistry
Chinese Academy of Sciences
Shanghai, China

Design of a δ opioid peptide mimetic based on a topographically constrained cyclic enkephalin

Victor J. Hruby [a], S. Liao [a], M. Shenderovich [a], J. Alfaro-Lopez [a], P. Davis [b], K. Hosohata [b], F. Porreca [b] and H.I. Yamamura [b]

Departments of [a]Chemistry and [b]Pharmacology, University of Arizona, Tucson, AZ 85721, USA

Introduction

A major goal in modern drug research is the ability to design *de novo* a non-peptide mimetic of the pharmacophore of a conformationally constrained receptor selective peptide hormone or neurotransmitter agonist for a 7-transmembrane G-linked receptor with high potency, selectivity and agonist efficacy. Thus far success has been very limited, since there is still much to learn about the structure, conformation, topography, and dynamics related to agonist activity at receptors [1]. For example, it is not clear that any of the many putative non-peptide structure mimetics or scaffolds [e.g. 2, 3] for peptide secondary structure (β-sheets, β-turns, etc.) actually can mimic the specific stereoelectronic, topographic and dynamic (e.g. α-helix-coil-β-sheet interconversion) properties of a peptide that may be critical in their agonist biological activities. On the other hand, a great deal of success has been realized in designing peptidomimetics that, although greatly modified structurally and topographically from the natural peptide agonist ligand, nonetheless retain many of the stereoelectronic and functional groups of peptides [e.g. 4, 5], as well as their agonist activities. Indeed, it is not clear what kinds of non-peptide scaffolds can be used to design ligands that can mimic the secondary structure, topography, and stereoelectronic properties of a peptide necessary for agonist activity. To help evaluate which conformationally flexible scaffolds may be used to mimic peptide agonist pharmacophores, we have used a highly potent, receptor selective, conformationally and topographically constrained peptidomimetic, computational chemistry and molecular modeling to design a non-peptide ligand that could mimic the major proposed pharmacophore elements. Based on these studies we have designed and synthesized a non-peptide mimetic of a δ receptor selective opioid ligand with high potency and selectivity similar to the peptide ligand on which it was based, but with low efficacy.

Results and Discussion

Our starting point for the design of a δ opioid receptor selective ligand was the cyclic enkephalin analogues c [DPen^2, DPen^5]enkephalin (DPDPE) for which we have the x-ray crystal structure [6], and the more receptor selective, topographically constrained peptidomimetic [(2S, 3R) β-methyl-2', 6'-dimethyltyrosine^1]DPDPE ([TMT^1]DPDPE), for which we have determined the conformation and preferred side chain topography for the (2S,3R)TMT[1] residue ($\chi^1 \cong 180°$; $\chi^2 \cong 80°$) and for the Phe[4] residue ($\chi^1 \cong -80°$; $\chi^2 \cong 90°$) [7]. Detailed structure-activity studies of the cyclic analogues of DPDPE, deltorphins and other potent, δ opioid receptor selective ligands demonstrated that the Tyr[1] phenol ring, the phenyl ring of the Phe[4] residue, and the α-amino group of the Tyr[1] residue were key pharmacophore elements for δ opioid receptor ligand affinity and transduction. Furthermore, δ opioid vs μ opioid receptor selectivity required the sterically bulky, lipophilic residues D-Pen[2](β, β-dimethylcysteine) and D(L)-Pen[5]. The three dimensional topographical features of all of these structural moieties were considered in our design (fig. 1). The piperazine scaffold we initially chose had several structural and conformational features that seemed to provide easy synthetic access to a variety of substituted ligands, and at the same time provide desirable topographical structural and stereoelectronic properties, plus considerable conformational flexibility [8]. Also of great importance in our design was the introduction of a large, bulky lipophilic R group to mimic the lipophilic DPen[2], D(L)Pen[5] methyl groups which were found to be so important for δ opioid receptor selectivity in cyclic enkephalins (fig. 1).

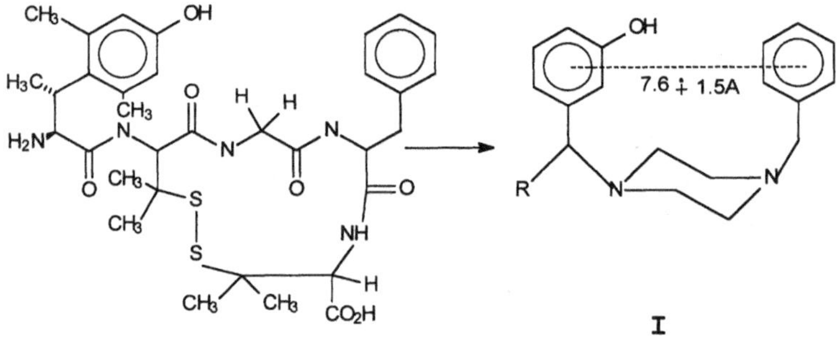

I

[(2S,3R)TMT[1]]DPDPE

*Fig. 1.Design of a first generation non-peptide mimetic based on the δ-opioid ligand [(2S, 3R)TMT[1]]DPDPE, in which the topographical relationships between the two aromatic moieties of (2S,3R)TMT[1] and Phe[4] are mimicked and the lipophilic bulky moiety is provided by R of **I** (see text).*

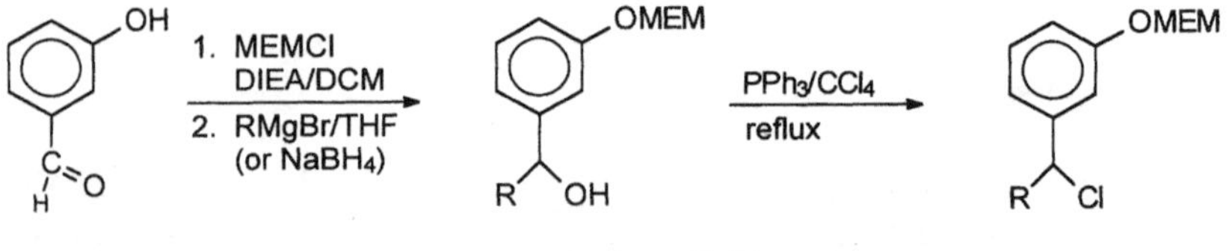

II

*Fig. 2. Synthesis of key intermediates (**II**) for the synthesis of non-peptide mimetic **I** with various R groups.*

For the synthesis of various analogues of I (fig. 1), a synthetic route to compounds II (fig. 2) was developed. The hydroxyl group of the starting material, 3-hydroxybenzaldehyde, was first protected and then the aldehyde either reduced (R=H) or treated with a Grignard reagent R = i-Bu or R = t-BuPh. The alcohol was converted to the halide II as shown in Fig. 2. Compounds II could be converted to analogues of I by condensation with N-benzylpiperazine, followed by removal of the protecting group to give I with various R-groups (table 1).

The synthetic non-peptide ligands were evaluated for their binding affinities using standard binding assay conditions developed in our laboratory against radiolabeled δ- and μ-receptor selective opioid ligands (table 1). As expected from our design, the binding affinities of **2**, **3** and **4** were quite different depending upon the structure of R. When R = H, binding affinity at both μ and δ opioid receptors is weak and there is little receptor selectivity (table 1). When R is a bulky aliphatic group such as iso-butyl (**3**, table 1) affinity for the δ receptor is increased about 15 fold at the δ opioid receptor, but only 4 fold at the μ receptor, leading to a slightly selective ligand (table 1). However, when R is the very bulky *p*-tert-butylphenyl group a potent (EC_{50} = 8.4) and δ opioid receptor selective (2000 fold) ligand is obtained (**4**, table 1). In fact this non-peptide mimetic of **1** is nearly as potent in affinity and is more selective than the peptide, [(2S, 3R) TMT1]DPDPE, on which its design was based.

A major question arises as to whether these non-peptide mimetics more closely mimic peptide or non-peptide ligands in their structure-activity relationships. For example, it has been shown that the hydroxyl group of the Tyr[1] residue is important for binding and transduction of peptide ligands with δ opioid receptors [9], whereas in non-peptide ligands such as (+)-BW373U8, methylation of the critical -OH group produces a highly selective and quite potent ligand SNC-80 [10]. Methylation of the phenol -OH group in DPDPE led to **5** which showed a 150-fold decrease in binding affinity at the δ opioid receptor (table 1) relative to DPDPE. Methylation of the -OH group in non-peptide **4** to give compound **7** led to a 210 fold decrease in affinity for the δ opioid receptor and to a compound with little δ vs μ receptor selectivity. Thus the peptide mimetics we have designed behave more like the peptide ligand DPDPE than non-peptide ligands such as SNC-80 and (+)-BW373U86.

Furthermore, it recently was reported [11] that potent and highly δ-opioid receptor selective peptide ligands and potent, selective non-peptide δ ligands interact differently with the wild type human compared to the mutated human δ opioid receptor in which the Trp284 residue was mutated to a Leu residue. In the peptide, no shift in the dose-response curve was seen when binding to the wild type and mutated receptors was compared, whereas for SNC-80 there was a 17-fold reduction in binding to the mutated receptor vs. the wild type. When we examined our designed non-peptide δ ligand **4** in the same binding assays under the same conditions, there was a 4-fold difference in binding was seen that was not statistically significantly from the parent compound.

Thus, our designed ligand behaved more like the peptide and displayed the same structure-activity relationships as the peptides, compared to the non-peptide ligands.

Table 1. Binding affinities of δ-Selective opioid ligands.

Compound	IC_{50} (nM)+SEM		Selectivity (μ/δ)
	vs [^{3}H]DAMGO(μ)	vs [^{3}H]pCl-DPDPE	
1. [(2S,3R)TMT^1]DPDPE[a]	4300 ± 820	5.0 ± 0.1	860
2. I-R = H	8100 ± 790	6400 ± 3200	1.3
3. I-R = i-Bu	2100 ± 600	420 ± 38	5.0
4. I-R = t-BuPh	17,000 ± 3000	8.4 ± 1.6	2000
5. [Tyr(OMe)^1]DPDPE	11,000 ± 2700	230 ± 24	48
6. DPDPE[a]	610	1.6	380
7. 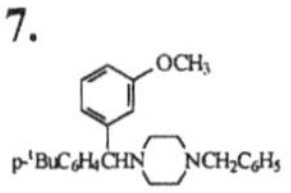	>8000	1800	>4

[a] see reference [7]

When **4** was tested in the MVD (δ receptor predominates) and the GPI (μ receptor predominates) bioassays, the potency of **4** dropped significantly to an EC_{50} of 85 nM and the compound was found to be a partial agonist relative to DPDPE and SNC-80 in the GTP (γS) assay (data not shown). Hence, despite its excellent binding affinity to δ opioid receptors and high selectivity at δ vs. μ receptors, **4** still does not transduce the opioid message as well as the compounds DPDPE and SNC-80. Although a potent and selective non-peptide δ-opioid receptor agonist analogue has for the first time been designed based on a peptide pharmacophore for that receptor, the best ligand (**4**, table 1) still appears to be deficient in transduction of the biological message through the receptor despite its high binding affinity and δ opioid receptor selectivity. The reason(s) for this are currently under investigation with further studies of the ligand **4** itself, and efforts are being made to further modify **4** so that it has more efficacious biological activity. Clearly the challenge remains to translate a potent and selective peptide agonist ligand pharmacophore for a 7-transmembrane G-linked receptor to a non-peptide ligand which is as potent, receptor selective, and efficacious as the native peptide.

Acknowledgments

This research was supported by grants from NIDA DA08657, DA06284 and DA04248. The views expressed here are those of the authors and not of the sponsoring agencies.

References

1. Hruby, V.J., Drug Discovery Today, 2 (1997) 165.
2. Giannis, A. and Kolter, T., Angew. Chem. Int. Ed. Eng., 32 (1993) 1244.
3. Sawyer, T.K., In Veerapandiam, V. (Ed.) Structure-Based Drug Design: Diseases, Targets, Techniques and Developments, Marcel Dekker, New York, 1997, p.559.
4. Rizo, J. and Gierasch, L.M., Ann. Rev. Biochem., 61 (1992) 387.
5. Hruby, V.J., Al-Obeidi, F. and Kazmierski, W.M., Biochem. J., 268 (1990) 249.
6. Flippen-Anderson, J.L., Hruby, V.J., Collins, N., George, C. and Cudney, B., J. Am. Chem. Soc., 116 (1994) 7523.
7. Qian, X., Shenderovich, M.D., Kövér, K.E., Davis, P., Horvath, R., Zalewska, T., Yamamura, H.I., Porreca, F. and Hruby, V.J., J. Am. Chem. Soc., 118 (1996) 7280.
8. (Shenderovich, M.D., Liao, S., Qian, X. and Hruby, V.J., In Tam, J.P. and Kaumaya, P.T.P. (Eds.) Peptide: Chemistry, Structure and Biology (Proceedings of the Fifteenth American Peptide Symposium), ESCOM Kluwer Acad. Publ., Dordrecht, 1998, unpublished.)
9. Hruby, V.J. and Gehrig, C.A., Med. Res. Revs., 9 (1989) 343.
10. Calderon, S.N., Rothman, R.B., Porreca, F., Flippen-Anderson, J.L., McNutt, R.W., Xu, H., Smith, L.E., Bilsky, E.J., Davis, P. and Rice, K.C., J. Med. Chem., 37 (1994) 2125.
11. Varga, E.V., Li, X., Stropova, D., Zalewska, T., Landsman, R.S., Knapp, R.J., Malatynska, E., Kawai, K., Mizusura, A., Nagase, H., Calderon, S.N., Rice, K., Hruby, V.J., Roeske, W.R. and Yamamura, H.I., Mol. Pharm., 50 (1996) 1619.

Tetrazole *cis*-amide bond mimetics identify the β-turn conformation of insect kinin neuropeptides

Ronald J. Nachman [a*], Janusz Zabrocki [b], Victoria A. Roberts [c] and Geoffrey M. Coast [d]

[a]*VERU, FAPRL, U.S. Department of Agriculture, 2881 F&B Rd., College Station, TX 77845, USA;* [b]*Institute of Organic Chemistry, Technical University, 90-924 Lodz, Poland;* [c]*Department of Molecular Biology, Scripps Research Institute, La Jolla, CA 92037, USA;* [d]*Department of Biology, Birkbeck College, University of London, London WC1E 7HX, UK,*
**To whom correspondence should be addressed.*

Introduction

The insect kinin neuropeptide family shares the common C-terminal pentapeptide Phe-Xaa^1-Xaa^2-Trp-Gly-NH_2 (Xaa^1 = Asn, His, Phe, Ser or Tyr; Xaa^2 = Ser, Pro or Ala), which represents the active core for diuretic activity in a variety of insects. They have been isolated from such diverse insect sources as the cockroach *Leucophaea maderae*, cricket *Acheta domesticus*, locust *Locusta migratoria*, corn earworm *Helicoverpa zea*, and mosquitoes *Culex salinarius* and *Aedes aegypti* [1, 2]. These kinins demonstrate Ca^{2+} ion dependent stimulation of Malpighian tubule fluid secretion and synergize the diuretic effects of the larger cAMP-dependent CRF-related diuretic hormones in insects [2]. The insect kinins at 1 nM double the rate of fluid secretion in cricket Malpighian tubules. This peptide family may therefore regulate water and ion balance in insects [1]. The development of peptidomimetic agonists/antagonists with enhanced resistance to degradative peptidases would provide important tools to neuroendocrinologists and could provide the basis for novel, selective pest control strategies via disruption of water and ion balance in target insects. A determination of the receptor-bound, or active, conformation would greatly aid the design of selective peptidomimetic analogs. A combined experimental and theoretical analysis of the aqueous solution conformation of the active conformationally restricted, cyclic insect kinin analog *cyclo*[Ala-Phe-Phe-Pro-Trp-Gly] indicated the presence of two major turn types; one over C-terminal pentapeptide core residues 1-4 (*cis*Pro) and the other over residues 2-5 (*trans*Pro) in a population distribution ratio of 60:40, respectively [3, 4] (fig. 1). Although the former conformation appeared to have a slight preference in solution and proved most consistent with structure-activity relationships inferred from the biological evaluation of a series of substitution analogs, the latter could not be ruled out as the conformation adopted by insect kinins at the receptor site. In this study, we synthesized analogs containing the tetrazole moiety, a *cis*-amide bond surrogate [5], inserted between active core residues Xaa^1 and Xaa^2. They were evaluated them in the Malpighian tubule fluid secretion assay to provide definitive evidence that the turn over residues 1-4 is the active conformation adopted by insect kinins at the receptor site.

Results and Discussion

Analysis of the active conformationally restricted, end-to-end cyclic insect kinin analog *cyclo*[Ala-Phe-Phe-Pro-Trp-Gly], in which distance and angle constraints obtained from aqueous solution NMR spectra were incorporated into molecular dynamics calculations, indicated the presence of two turn types over two distinct residue blocks within the C-terminal pentapeptide core region. These two conformations are illustrated in Fig. 1. The more populous of the two turns in aqueous solution features a *cis*Pro in the third position of a type VI β-turn over active core residues 1-4, or Phe-Phe-**Pro**-Trp. The second turn contains a *trans*Pro in the second position of a less rigid type-I-like β-turn over active core residues 2-5, or Phe-**Pro**-Trp-Gly. From molecular dynamics calculations, the most favorable *cis*Pro structure had an intramolecular energy about 7 kcal/mol lower than the most favorable *trans*Pro structure, suggesting that it was be the predominant conformation. Indeed, NMR spectra indicated that the *cis*Pro turn was the most populous conformation in aqueous solution by a slight 60:40 ratio [3,4]. This is in agreement with systematic studies on linear peptides with Pro^3 [6] in which flanking aromatic residues promote the formation of type VI β-turns in aqueous solution. Such turns are further enhanced when small, hydrophilic residues (i.e., Asp, Ser, Thr, **Gly** or Asn) follow the aromatic-Pro-aromatic motif [6]. The molecular modeling studies on the cyclic insect kinin analog indicate that interactions between the aromatic side chains in positions 1 and 4 help stabilize the turn over residues 1-4 containing the *cis*Pro configuration, which would otherwise be less energetically favorable than *trans*Pro [4].

An examination of the *cis*Pro structure (fig. 1a) indicates that the aromatic side chains of Phe^1 and Trp^4, both critical for biological activity, are adjacent to one another and oriented on the same side of the main chain backbone. In contrast, the side chain of residue 2 (Phe^2) lies on the opposite side of the main chain backbone, extending away from the surface formed by Phe^1 and Trp^4. Position two is variable and tolerates a wide variety of substitutions that do not occur naturally, ranging from basic to acidic, and from hydrophilic to hydrophobic, without complete loss of biological activity [3, 4]. In contrast, the two essential aromatic side chains are far apart and are on opposite sides of the cyclic peptide backbone of the *trans*Pro structure (fig. 1b). A plausible receptor-interaction model would place the insect kinins in a *cis*Pro conformation, the receptor leading with the critical Phe^1/Trp^4 aromatic surface, leaving the residue in the variable and tolerant position 2 pointing directly away from the receptor surface.

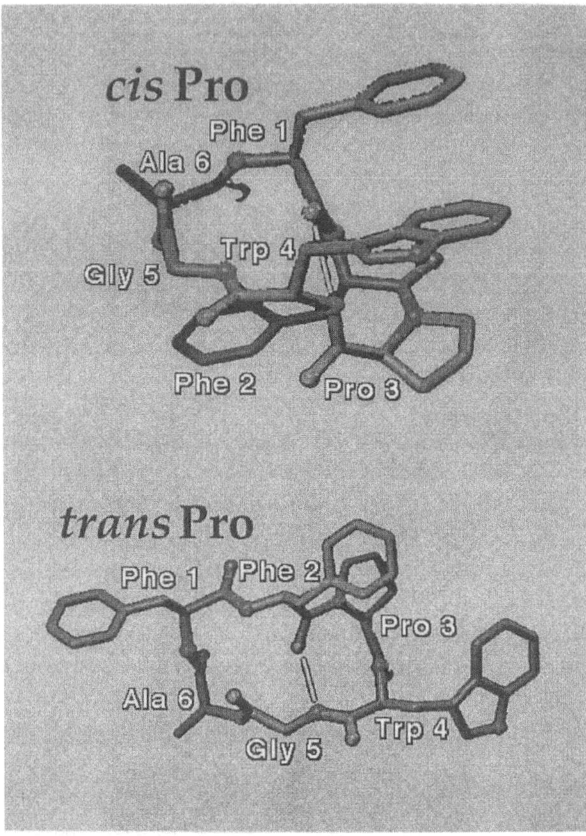

Fig. 1. Low energy conformations of cyclo[Ala-Phe-Phe-Pro-Trp-Gly] found by computational search on Convex and Cray XMP supercomputers with programs Discover and Insight [4]. (a) Above is a low-energy cisPro structure with a type VI turn over residues 1-4, in which the Phe^1 and Trp^4 side chains lie near each other on one side of the cyclic main chain. The sequence variable position 2 side chain (in this case Phe) lies on the opposite side of the ring, extending away from the essential 1 and 4 side chain surface. (b) Below is a low energy transPro structure with a type I turn over residues 2-5 in which side chains 1 and 4 diverge from each other.

In order to provide definitive evidence that the *cis*Pro type VI β-turn is the active insect kinin conformation, analogs containing a tetrazole link between active core residues Xaa^1 and Xaa^2 were synthesized to mimic the *cis*-amide bond type VI turn. The amide bond [-C(=O)N(H)-] is mimicked by a [-C(=N)N(N-)-] moiety and tied into a *cis* orientation by a nitrogen atom (-N=) in the cyclic tetrazole. The synthetic tetrazole turn mimic fragment-blocks Boc-Phe-Phe-ψ[CN_4]-Ala-OH, in L and D forms (with respect to the Ala α-carbon), was synthesized according to previously described general procedures [5]. The tetrazole fragment-blocks were incorporated onto the peptide-resin complex, Trp-Gly-MBHA, and cleavage was affected under previously described conditions [1] to afford Phe-Phe-ψ[CN_4]-Ala-Trp-Gly-NH_2 in L and D forms (fig. 2).

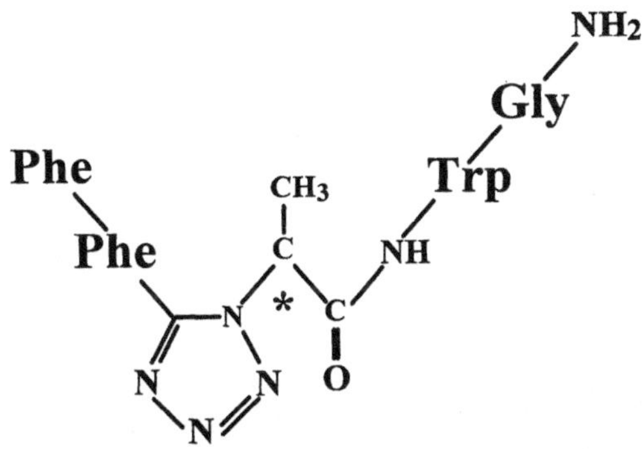

Fig. 2. Structure of tetrazole peptidomimetic analogs of the insect kinin C-terminal pentapeptide active core. Both L and D stereoisomers were synthesized with respect to the chiral center at the α-carbon of the Ala (denoted by an asterisk). The tetrazole represents a mimic of the cis amide peptide bond configuration [5]. In this figure, the imide at the left side of the tetrazole ring replaces the amide bond carbonyl of Phe^2 and the NH of Trp^4 is illustrated to the left of its three-letter code. The analogs mimic a cisPro type VI turn over residues 1-4 of the natural peptides.

Evaluation in a cricket Malpighian tubule fluid secretion bioassay demonstrated that the L version of the insect kinin tetrazole analog retained significant diuretic activity with an EC_{50} of 0.34 μM (95% CL 0.27 - 0.43 μM) and matches (106%) the maximal response of native insect kinins. In contrast, the D tetrazole analog exhibited unusual behavior with partial antagonist activity, reducing the diuretic response of a 1 nM solution of the natural cricket achetakinin-1 (AK-1) by 50% at an IC_{50} of 0.44 μM (95% CL 0.34 - 0.56) (fig. 3). It also showed partial agonist activity with only a 50% maximal response and an EC_{50} of 0.19 μM (G.M. Coast and R.J. Nachman, unpublished). Although the L tetrazole analog demonstrated reduced potency it matched the efficacy of the native insect kinins, suggesting that it interacted successfully with the tubule receptor site despite its inherent rigidity. Furthermore, the D analog appears to be able to bind with the receptor while its ability to activate the receptor is diminished, leading to the observed partial antagonist behavior. The results, in aggregate, provide strong evidence that the active conformation adopted by the insect kinins at the Malpighian tubule receptor site is the *cis*Pro type VI β-turn over residues 1-4. This information will be invaluable in the design of selective peptidomimetic agonists/antagonists. The tetrazole analogs themselves represent a lead template from which more potent agonist and/or antagonist analogs can be developed through the use of a combinatorial peptide library approach [7, 8]. The presence of the tetrazole linkage in the insect kinin active core region should offer a measure of protection from degradation, as analogs are likely to be resistant to peptidases that target cleavage sites in the vicinity of this moiety. Disruption of the water and/or ion balance of target insects with peptidomimetic insect kinin analogs with enhanced resistance to hemolymph and tissue-bound peptidases could form the basis for novel, selective pest management strategies in the future [1].

Acknowledgments

NATO grant No. 90248 (R.J.N. and G.M.C.)

References

1. Nachman, R.J., Isaac, R.E., Coast, G.M. and Holman, G.M., Peptides, 18 (1997) 53.
2. Coast, G.M., In Coast, G.M. and Webster, S.G. (Eds.) Recent Advances in Arthropod Endocrinology (Society for Experimental Biology Seminar Series 65), Cambridge University Press, Cambridge, UK, 1998, p. 189.
3. Nachman, R.J., Roberts, V.A., Holman, G.M. and Tainer, J.A., In Epple, A., Scanes, C.G. and Stetson, M.H. (Eds.) Progress in Comparative Endocrinology, Wiley-Liss, Inc., New York, NY, 1990, p. 624.
4. Roberts, V.A., Nachman, R.J., Coast, G.M., Hariharan, M., Chung, J.S., Holman, G.M., Williams, H. and Tainer, J.A., Chemistry & Biology, 4 (1997) 105.
5. Zabrocki, J., Dunbar, J.B., Marshall, K.W., Toth, M.V. and Marshall, G.B., J. Org. Chem., 57 (1992) 202.
6. Yao, J., Feher, V.A., Espejo, B.F., Reymond, M.T., Wright, P.E. and Dyson, H.J., J. Mol. Biol., 243 (1994) 736.
7. Furka, A., Int. J. Peptide Protein Res., 37 (1991) 487.
8. Houghton, R.A., Biopolymers, 37 (1995) 221.

Synthetic studies of farnesyl protein transferase inhibitor pepticinnamin E

Jin-Gen Deng, De-Qun Sun, Xiao-Qin Lin, Ai-Qiao Mi and Yao-Zhong Jiang

Union Laboratory of Asymmetric Synthesis, Chengdu Institute of Organic Chemistry, Chinese Academy of Sciences, P.O. Box 415, Chengdu, 610041, China

Introduction

Pepticinnamin E (**1**) is a major product of the pepticinnamins isolated from the culture of *Streptomyces* sp. OH-4652. It has been identified as a depsipeptide containing o-Z-pentenylcinnamic acid (**2**) and a novel dopa analog (**3**), whose absolute configuration has not been determined. It showed rather potent inhibitory activity against farnesyl protein transferase (FPTase) with an IC_{50}, of 0.3 μM and is the first competitive inhibitor derived from a natural product of the substrate p21 ras protein with *Ki*, 1.76 μM [1]. Our interests in the exploitation of new methodology to synthesize natural products containing non-ribosomal amino acids [2] and studies of antitumor activity against ras-dependent tumor led us to initiate the synthesis of pepticinnamin E (**1**). Herein, we report our progress in synthesizing pepticinnamin E (**1**).

Results and Discussion

We began with the synthesis of o-Z-pentenylcinnamic acid (**2**), in which the key step is a *Z*-selective Wittig's reaction of 2-carboxybenzaldehyde (**6**). Generally, nonstabilized ylides predominantly yield Z-alkene [3]. Ludfk and co-workers reported that more than 96% *Z*-selectivity was obtained via a Wittig's reaction of *p*-methylbenzaldehyde with lithium 1,3-diaminopropane and pentyltriphenylphosphorane bromide [4]. However, o-*E*-pentenylbenzoic acid (**7**, *Z/E*=33:67) was predominantly produced in the reaction of aldehyde (**6**) and butyltriphenylphosphorane bromide under the Ludfk condition and, interestingly, a high *Z*-selectivity (*Z/E*, 98:2) of (**7**) was obtained under the Still condition [5]. Thus, ethyl o-Z-pentenylcinnamate (**9**) was given following reduction, oxidation and Wittig's reaction. Unfortunately, the instability and the limits on large-scale preparation of (**7**) urged us to investigate a new methodology to synthesize o-Z-pentenylcinnamic acid (**2**). The key Heck reaction quantitatively gave (**11**), followed by oxidation to yield aldehyde (**12**). The Wittig's reaction with the phosphorane and butyllithium then gave a high *E*-selective product (**13**), which was treated with the bases to provide arylacetylene (**14**) with 35%~41% yield. Alkylation of (**14**) was investigated under different conditions, and was ultimately considered to be the steric effect of the *tert*-butoxycarbonylvinyl group. Thus, the hydroxyl group of 2-iodobenzyl alcohol (**10**) was protected with THP. THP-protected arylacetylene (**17**) was synthesized through the important Heck reaction catalyzed by palladium with 2-methyl-3-butyn-2-ol and subsequent removal of acetone with sodium hydride. Furthermore, the propylation of (**17**), followed by Wittig's reaction and cis-selective hydrogenation on Lindlar catalyst was expected to give desired product, o-*Z*-pentenylcinnamic acid (**2**).

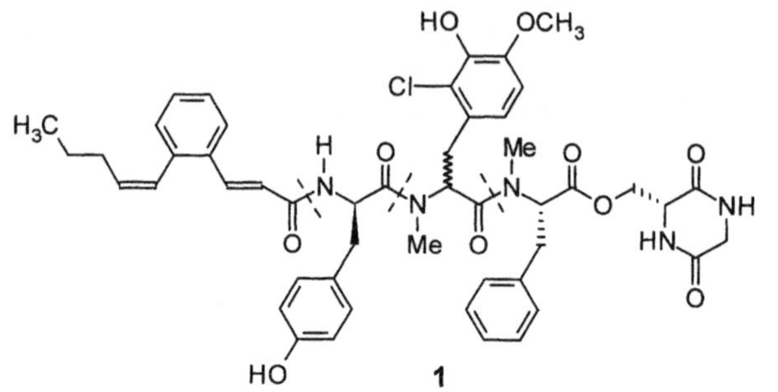

1

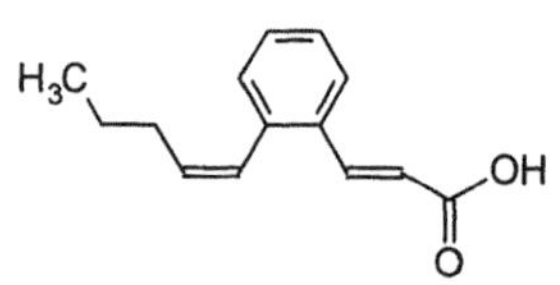

2

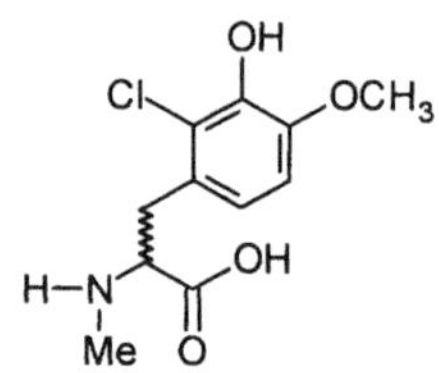

3

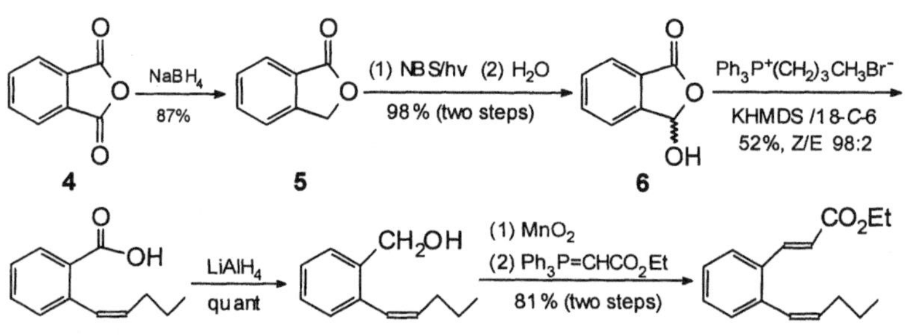

4 **5** **6**

7 Z-isomer **8** **9**

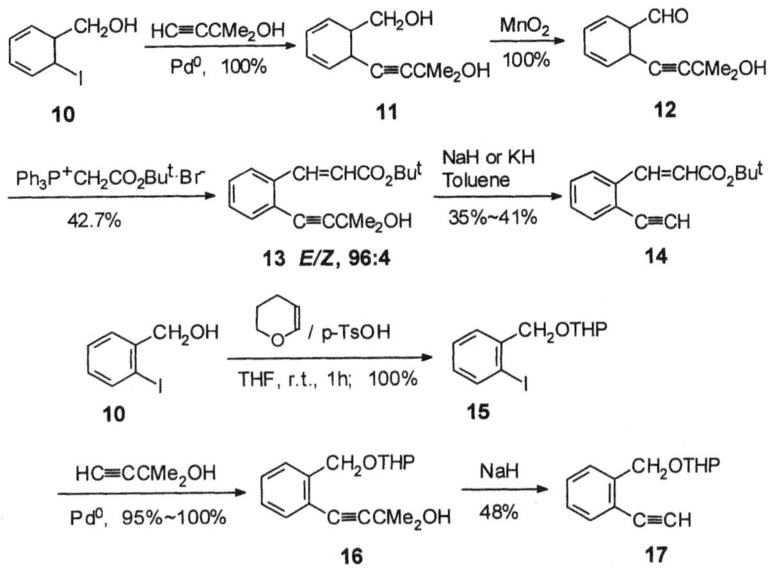

References

1. Omura, S. and Tomoda, H., Pure & Appl. Chem., 66 (1994) 2267.
2. Deng, J. G., Hamada, Y. and Shioiri, T., Tetrahedron Lett., 37 (1996) 2261.
3. Yamataka, H., Nagareda, N., Ando, K. and Hanafusa, T., J. Org. Chem., 57 (1992) 2865.
4. Ludfk, S., David, S., Karel, K., Jam, V. and Miroslav, R., CS 274,242 (Cl. C07C1/34) (1992).
5. Still, W. C. and Gennari, C., Tetrahedron Lett., 24 (1983) 4405.

Novel coupling reagents

Peng Li and Jie-Cheng Xu*
Shanghai Institute of Organic Chemistry, Chinese Academy of Sciences, shanghai, 200032, China

Introduction

In recent years, many new coupling reagents have been suggested for replacing conventional methods for peptide synthesis to enhance coupling efficiency and eliminate racemization. We have developed some new high-reacting coupling reagents; which are mainly uronium salts or phosphinates [1-4]. We report here the synthesis of some immonium salts and evaluate their efficiency in coupling peptides by comparisons with some popular reagents.

Results and Discussion

The immonium salts (1-3) were readily prepared from the condensation of DMF with BTC [bis(trichloromethyl)carbonate] or $POCl_3$ to yield the intermediate, immonium chlorides, which were subsequently reacted with the corresponding hydroxyl-containing compounds and stabilized with the ionic counterpart, $SbCl_5$, to give the desired compounds 1-3.

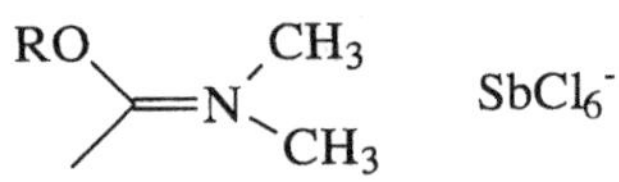

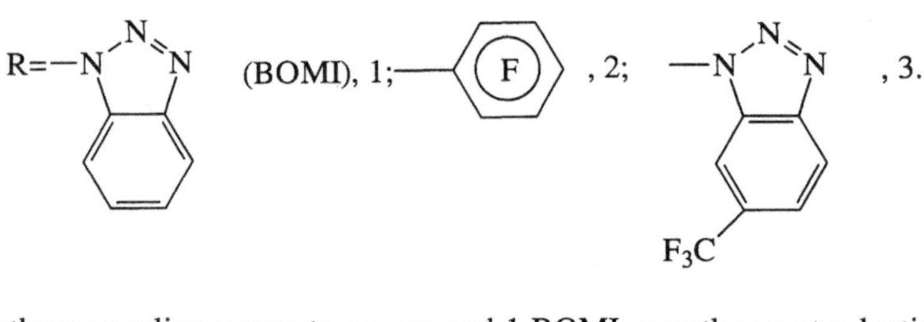

Among these coupling reagents, compound 1 BOMI, was the most selective because of its convenience of preparation and fair stability in storage. Its efficiency in oligopeptide synthesis, the racemization estimates and the influence of several reaction parameters such as solvent, base, and temperature were studied by HPLC method using the following model reaction:

Z-Gly-Phe-OH+Val+OMe·HCl→Z-Gluy-(D+L)PHe-Val-Ome, in which phenylalanine is sensitive to stereomuation, especially in apolar solvents [5], resulted in two diastereometric tripeptides (Z-Gly-L-Phe-Val-OMe and Z-Gluy-D-Phe-Val-OMe) which can be separated easily and monitored at 220 nm due to the absorbance of the amide bond. A low level of racemization for the above model coupling (1.55-2.4%) was observed using HPLC method, while 4.4% racemization was otherwise observed by Young's test during coupling of Bz-Lue-OH with Gly-Oet·HCl.

Several oligopeptides were prepared using BOMI for assessing effectiveness (table 1). The reaction was carried out as usual as HBPyU. 2,6-lutidine was added to a cold mixture (-10 °C) of N-protected amino acid (1 equiv.), amino acid ester hydrochloride (1.1 equiv.), and coupling reagent: BOMI (1.1 equiv.) in CH_3CN (1 ml/mmol), and stirred for 1 minute in the cold and for 15 min at room temperature. TLC monitoring showed that the complete conversion was reached within only a few minutes.

Table 1. Prepare of peptides with BOMI in solution.

Entry	Peptide(yield %)[a]	m. p.(°C)	$[\alpha]_D$(conc., solv.)
1	Boc-Phe-Gly-Oet[b](91)	98-100	-5.3 (2, EtOH)
2	Boc-Phe-Gly-OMe[c](91)	116-117	-14 (1, EtOH)
3	Z-Ser-Gly-OEtd(88)	107-108	-6.4 (1, AcOEt)
4	Boc-Ile-Val-Ome[e](79)	165-166	-15.5 (1, AcOEt)
5	Bz-Leu-Gly-Oetf(84)	154-155	-31(3.1, etOH)

a) Yield based on the amount of N-protected amino acid used
b) Lit[6]: mp=88-89.5 °C; $[\alpha]_D^{20}$ =-4.2 (2, EtOH)
c) Lit[7]: mp=114-115 °C; $[\alpha]_D^{20}$ =-14 (1, EtOH)
d) Lit[8]: mp=106-107 °C; $[\alpha]_D^{20}$ =-5.9 (1, AcOEt)
e) Lit[9]: mp=166-167 °C; $[\alpha]_D^{20}$ =-15 (1, AcOEt)
f) Young's test [10] Lit[10]; mp=166-167 °C; $[\alpha]_D^{20}$ =-15 (1, AcOEt)

The factors affecting coupling, such as solvent, base and temperature were examined. More detailed results will be published elsewhere. To conclude, a new type of coupling reagents the immonium salts, particularly BOMI, was shown to be a very promising peptide coupling reagent in terms of the moderate to good yield, fast conversion rate and low racemization.

Acknowledgments

This work was supported by the National Natural Science Foundation of China.

References

1. Chen, S.Q. and Xu, J.C., Tetrahedron Lett., 32 (1991) 6711.
2. Chen, S.Q. and Xu, J.C., Tetrahedron Lett., 33 (1992) 647.
3. Chen, S.Q. and Xu, J.C., Chinese Chemical Letters, 4 (1993) 847.
4. Chen, S.Q. and Xu, J.C., Chinese Journal of Chemistry, 13 (1995) 175.
5. Van der Auwera, C., Van Damme, S. and Anteunis, M. J. O., Int. J. Pept. Protein Res., 29 (1987) 464.

6. Paul, R. and Anderson, G. W., J. Amer. Chem. Soc., 82 (1960) 4596.
7. Nitecki, D. E., Halpern B. and Westley, J. W., J. Org. Chem., 33 (1968) 864.
8. Yamada S. and Takeuchi, Y., Tetrahedron Letters, 39 (1971) 3595.
9. Coste, J., Le-Nguyen D. and Castro, B., Tetrahedron Lett., 31 (1990) 205.
10. Williams, M. W. and Young, G. T., J. Chem. Soc., (1963) 881.

Pharmacophore model of growth hormone secretagogues

Liang Liu, Ren-Xiao Wang, Lu-Hua Lai and Chong-Xi Li
Chemistry Department, Peking University, Beijing, 100871, China

Introduction

Growth hormone releasing peptide (GHRP) is the artificial exogenous peptide hormone that stimulates growth hormone secretion. GHRPs and their peptidomimetics are currently being evaluated clinically as alternatives to GH replacement therapy in the treatment of growth hormone deficient children [1].

In this paper, we selected 105 non-peptide GH secretagogues [2-13] and established a pharmacophore model using distance-comparisons (DISCO) and comparative molecular field analysis (CoMFA) [14, 15]. These compounds are analogues of L-692, 429, MK-0677 and G-7502 (fig. 1). All modeling was performed with the program Sybyl6.3, run on Silicon Graphics workstation.

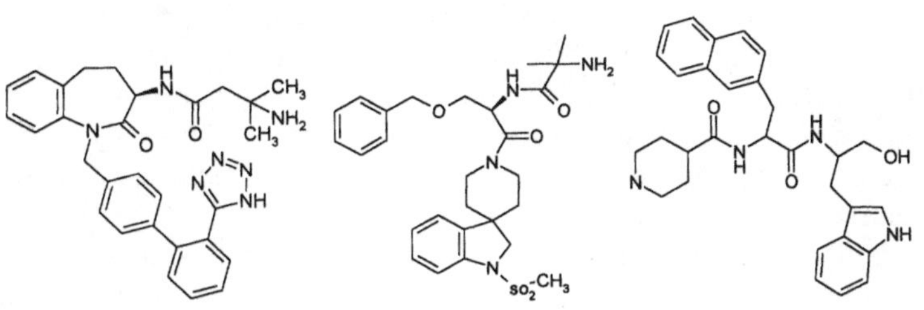

L-692, 429 MK-0, 677 G-7502

Fig.1. Typical structures of GH secretagogues.

Results and Discussion

DISCO

16 compounds with high bioactivity were chosen to derive the pharmacophore model by the distance-comparison method. Systematic conformational search was performed on each molecule to obtain low energy conformations. These conformations were then minimized using the conjugate-gradient method. The 3 classes of features in DISCO calculation were H-bond donor atoms, H-bond acceptor atoms and hydrophobic centers. One of the most active among these compounds was chosen as the reference compound. DISCO found one molecule at a tolerance of 2.39 Å that included 3 H-bond acceptor atoms, 2 H-bond donor atoms and 2 hydrophobic centers. Fig. 2 shows the superimposition of L-692, 429 and the reference compound.

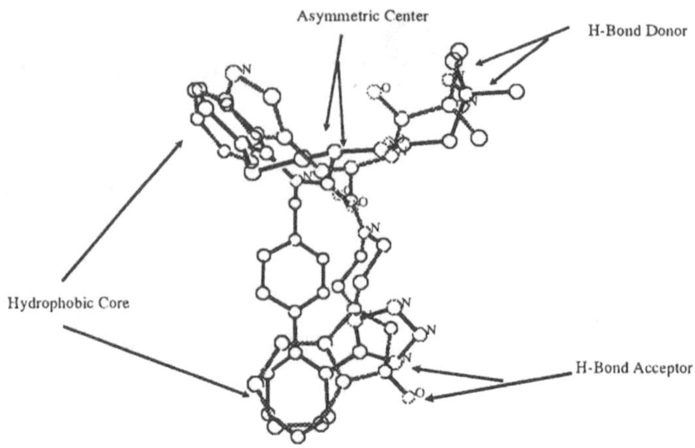

Fig. 2. The overlay of L-692, 429 and the reference molecule. The dark molecule is L-692, 429, the light one is the reference molecule.

CoMFA

From the ensemble of 100 GH secretagogues, a training set consisting of 62 structures and a test set of 38 compounds were randomly chosen. The computer model of all molecules except those in the DISCO model were generated by modifying the template molecule (the most similar structure in model). Systematic conformational search was performed on the side chain of each molecule. After minimiziation by using the Powell method, these conformations were fitted on the template molecule and saved in new molecular databases.

To improve the predictive ability of QSAR model, we developed a new CoMFA method. In this method, all the conformations of one molecule in the training set were treated as candidates, and the one that made the best contribution to the conventional CoMFA result was selected and added in the final model. This method therefore was called conformation choice-comparative molecular field analysis (CC-CoMFA). The default setting was used for all steps of conventional CoMFA.

The cross-validated r^2 (r^2_{cv}) of the training set is 0.695 with 5 components and the predictive r^2 (r^2_{pred}) value is 0.933. Fig. 3 (A) shows the actual-versus-predicted plot of the training set. The predictive power of this CoMFA model was also tested by predicting the activity of molecules in the test set. The r^2_{pred} of the test set is 0.746 and the actual-versus-predicted plot is shown in Fig. 3 (B).

These results proved that the CoMFA model of GH secretagogues is reliable and that the new method we applied in the modeling could recognize the best conformation alteration of the ligand. This procedure improved the predictive ability and reliability of the CoMFA model, which has been applied in the design of new growth hormone secretagogues.

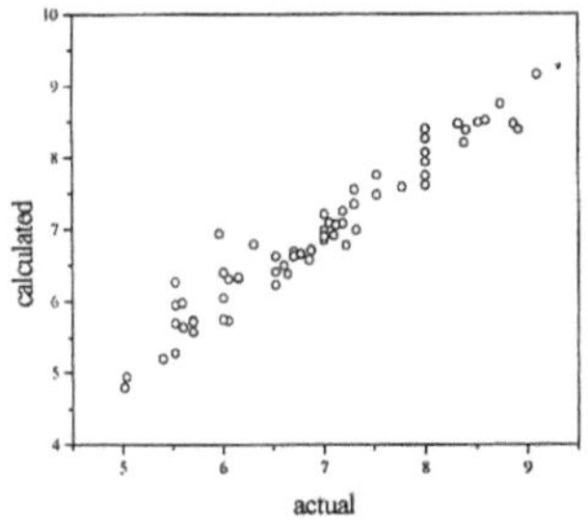

(A)

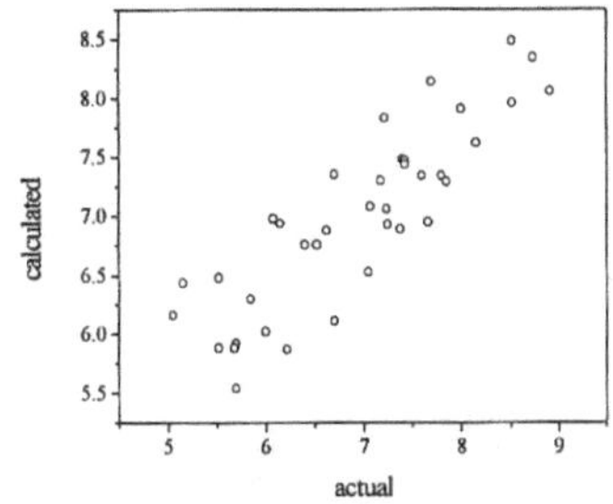

(B)

Fig. 3. Actual versus predicted plot for the training set(A) and the test set(B).

References

1. Korbonits, M., Trends Endocrinol Metab, 6 (1995) 43
2. William, R., Bioorg & Med Chem. Lett., 4 (1994) 1117
3. DeVita, R., Bioorg & Med Chem. Lett., 4 (1994) 1807
4. DeVita, R., Bioorg & Med Chem. Lett., 4 (1994) 2249
5. Ok, D., Bioorg & Med Chem. Lett., 4 (1994) 2709
6. Nargund, R.P. Bioorg & Med Chem. Lett., 6 (1996) 1265
7. Nargund, R.P., Bioorg & Med Chem. Lett., 6 (1996) 1731
8. Chen, M.H., Bioorg & Med Chem. Lett., 6 (1996) 2163
9. Ok, H.O., Bioorg & Med Chem. Lett., 6 (1996) 3051
10. Tata, J.R., Bioorg & Med Chem. Lett., **7** (1997) 663
11. Ankersen, M., Bioorg & Med Chem. Lett., **7** (1997) 1293
12. Patchett, A.A., Proc. Natl. Acad. Sci. USA, 92 (1995) 7001
13. Yang, L., Bioorg & Med Chem. Lett., 8 (1998) 107
14. Martin, Y. C., J. Computer-Aided Molecular Design, **7** (1993) 83
15. Cramer III, R. D., J. Am. Chem. Soc., 110 (1988) 5959

Discovery that deltorphin II derivatives are potent melanotropins, putatively active at the *Xenopus* melanocortin-1 receptor

V.J. Hruby [a], G. Han [a], M.J. Quillan [b], W. Sadee [b] and S. Sharma [a]

[a]*Department of Chemistry, University of Arizona, Tucson, AZ 85721-0041;*
[b]*Department of Pharmaceutical Chemistry, UCSF, San Francisco, CA 94143, USA*

Introduction

α-Melanotropin stimulating hormone (α-MSH, Ac-Ser-Tyr-Ser-Met-Glu-His-Phe- Arg-Trp-Lys-Pro-Val-NH_2, **1**) is an important peptide hormone and neurotransmitter. It has been studied extensively for its important role in pigmentation [1] by interaction with the melanocortin 1 receptor (MC1R). Recently other melanocortin receptors, MC3R, MC4R and MC5R were discovered [2]. These new receptors appear to be involved in several critical biological functions, including cardiovascular [3, 4] and antipyretic activities[5]. Recent discoveries that α-MSH analogues modulate feeding behavior [6] and erectile function have attracted much attention worldwide. These findings have stimulated efforts to develop highly potent and selective receptor ligands to determine the physiological roles and pharmacological properties of these receptors and of naturally occurring or designed ligands. Several approaches have been developed in our laboratory to generate constrained ligands with high potency and selectivity by mimicking the β-turn core in MSHs [7-9]. However, there is a need to search for new ligands with different MCR profiles.

Our approach to design MSH analogues with new profiles originated from considerations of the proopiomelanocortin gene (POMC) of which α-MSH is a processed product. POMC also produces other MSH peptides as well as β–endorphin, an extensively studied opioid. This led us to speculate that there may be a relationship between opioids and MSHs. As an initial study, analogues **3-8** of Deltorphin II (H-Tyr-*D*Ala-Phe-Glu-Val-Val-Gly-NH_2 **2**), a δ-receptor selective opioid, (table 1) were chosen for examination.

Results and Discussion

By extending the N-terminus of Deltorphin II (**2**) with the basic residue lysine, of the EC_{50} derivative **7** was 21 nM (table 1). However, when the N-terminus of **2** was extended with the more basic residue, arginine, the EC_{50} of compound **8** decreased to about 1 nM. It seems that either an increase of basicity or hydrophilicity of the side chain may increase the potency of N-terminal modifications.

Table 1. Bioassay of deltorphin II analogues in Xenopus frog.

#	Structure	EC_{50} (nM)
3	⌊NH-Lys-Tyr-DAla-Phe-Glu⌋-Val-Val-Gly-NH_2 (cyclic, bridge between NH and Glu)	490 ± 78
4	H_2N-Lys-Tyr-DPen(SH)-Phe-Glu-Pen(SH)-Val-Gly-NH_2	319 ± 151
5	⌊HN-Arg-Tyr-DAla-Phe-Glu⌋-Val-Val-Gly-NH_2 (cyclic, bridge between HN and Glu)	265 ± 168
6	⌊HN-Lys-Arg-Tyr-DAla-Phe-Glu⌋-Val-Val-Gly-NH_2 (cyclic, bridge between HN and Glu)	35 ± 12
7	H_2N-Lys-Tyr-DAla-Phe-Glu-Val-Val-Gly-NH_2	21 ± 18
8	H_2N-Arg-Tyr-DAla-Phe-Glu-Val-Val-Gly-NH_2	~ 1

Compound **3** is a constrained version of **7** in which a lactam bridge is formed between the N-terminus and the side chain of glutamic acid. This restriction led to a derivative more than 20 fold less potent than **7**, with EC_{50} around 0.5 μM. Likewise, the same kind of restriction of **8** resulted in **5** with an EC_{50} of 0.3 μM and a more than 200 fold loss of potency. The significant loss of activity after restriction of parent peptides indicates that the constrained conformation interacts poorly with MCRs. Other pathways to constrain peptides should be considered in order to find analogues more potent and selective than the linear parent peptides.

On the other hand, we found that deltorphin derivative **4,** which is topographically restricted with *D*-Pen and Pen, which replaced *D*-Ala and Val, respectively, in compound **7**, suffered a 15-fold loss of activity. The active conformation of **7** apparently was altered by such constraint replacement. Consequently, active conformers of **4** interacted with the receptors much less effectively.

Interestingly, extending the N-terminus of 5 (20 member-ring) with a Lys residue to give **6** (23 member-ring), enhanced potency by almost one order of magnitude, although **6** is still about 30-fold less potent than **8**, the linear version of **5**.

In summary, a potent MSH analogue **8** with a basic N-terminal Arg residue derived from deltrophin II has been identified. This is the first demonstration that certain opioid analogues can interact effectively with the putative MC1R. In addition, we have found that certain restrictions of the designed deltorphin analogues did not give enhanced activities. More research is necessary to answer the many puzzles encountered in this research.

Acknowledgements

Financial support from USPHS Grant No. DK17420 is greatly appreciated.

References

1. Vaudry, H. and Eberle, A.N., Ann. New York Acad. Sci., 680 (1993) 1.
2. Cone, R.D., Mountjoy, K.G., Robbins, L.S., Nadeau, J.H., Johnson, K.R., Roselli-Rehfuss, L. and Mortrud, M.T., Ann. New York Acad. Sci., 680 (1993) 342.
3. Van Bergen, P., Kleijne, J.A., De Wildt, D.J. and Versteeg, D.H.G., Brit. J. Pharmacol, 120(1997) 1561.
4. Li, S.J., Varga, K., Archer, P., Hruby, V.J., Sharma, S.D., Kesterson, P.A., Cone, R.D. and Kunos, G., J. Neurosic., 16 (1996) 5182
5. Huang, Q.H., Entwistle, M.L., Alvaro, J.D., Duman, R.S., Hruby, V.J. and Tatro, J., J. Neurosci., 17 (1997) 3343.
6. Fan, W., Boston, B.A., Kesterson, R.A., Hruby, V.J. and Cone, R., Nature, 385 (1997) 165.
7. Sugg, E.E., Castrucci, A.M., Hadley, M.E., van Binst, G. and Hruby, V.J., Biochemistry, 27 (1988) 8181.
8. Hruby, V.J., Sharma, S.D., Toth, K., Jaw, J.Y., Al-Obeidi, F., Sawyer, T.K., Castrucci, A.M. and Hadley, M.E., Ann. New York Acad. Sci., 680 (1993) 51.
9. Hruby, V.J., Han, G. and Hadley, M.E., Letters Pept. Sci., 5 (1998) 1.

Session F
Glyco/Lipo/Phospho peptide; Peptide diversity/Chemical libraries; Receptor-ligand interaction

Chairs: Jean Martinez
CNRS Faculté de Pharmacie
France

Yun-Hua Ye
Department of Chemistry
Peking University
Beijing, China

Enzymatic synthesis of phosphopeptides

Yan-Mei Li [a], Yu-Fen Zhao [a] and Herbert Waldmann [b]
[a]*Department of Chemistry, Tsinghua University, Beijing, 100084, China;* [b]*Institut fuer Organische Chemie, Universitaet Karlsruhe, 76128 Karlsruhe, Germany*

Introduction

Reversible protein phosphorylation is widely recognized as an important mechanism for the regulation of many cellular processes [1]. For the study of the roles of the phosphoproteins in biological processes, structurally well-defined compounds, which contain the characteristic structural units of the parent biomacromolecules are often needed. To develope therapeutic agents that target intracellular phosphorylation and dephosphorylation processes, characteristic peptides incorporating the phosphorylation sites of their parent, naturally occurring phosphoproteins, serve as invaluable tools.

The synthesis of phosphopeptides and other polyfunctional compounds calls for the use of various orthogonally stable protecting groups that have structural complexity, as well as pronounced chemical lability. In this study, enzymatic protecting group techniques were chosen to synthesis those sensitive compounds.

Selection and synthesis of protecting group

The enzyme penicillin G acylase from *E. Coli* [2], which attacks phenylacetic acid amides but does not hydrolyze peptide bonds, has been applied for N-terminal phosphopeptide deprotection.

In this study, p-phenylacetoxybenzyloxycarbonyl (PhAcOZ) group was developed. This is a urethane which embodies a functional group (here: a phenylacetate) that is recognized by the biocatalyst (here: penicillin G acylase) and that is bound by an enzyme-labile linkage (here: an ester) to a functional group. Spontaneous fragmentation occurs upon cleavage of the enzyme-sensitive bond (here: a p-hydroxybenzyl urethane) resulting in the liberation of a carbamic acid derivative, which decarboxylates to give the desired phosphopeptide (scheme I).

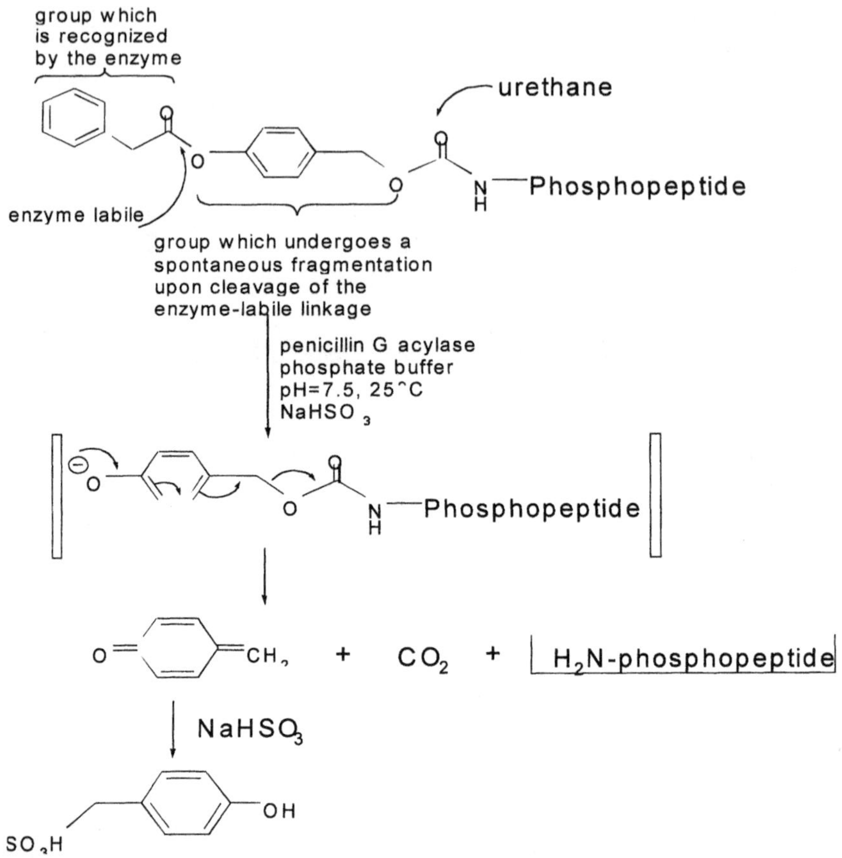

Scheme I.

For the C-terminal, the heptyl (Hep) was developed as a carboxyl protecting group, which can be removed by lipases (of the various lipases investigated, the enzyme from the fungus *Aspergillus niger* is the best.) (scheme II).

$$\text{PG-Phosphopeptide-OHep} \xrightarrow[\text{pH=7}]{\text{lipase from Aspergillus niger}} \text{PG-Phosphopeptide-OH}$$

10 vol% aceton

Scheme II.

Phosphorylation of peptides or amino acids

Reversible protein phosphorylation is widely recognized as an important mechanism for the regulation of many cellular processes. The synthesis of model phosphopeptide compounds is very important in the study of their biological phenomena.

First, peptides or amino acids were phosphorylated by dialkyl phosphoric chloride. However, this failed. It is successful when bis-alkyloxy-N, N-dialkylphosphoramidit is used followed by oxidation. Furthermore, many hydroxyl groups in serines or threonines can be phosphorylated in one step.

Typical example:

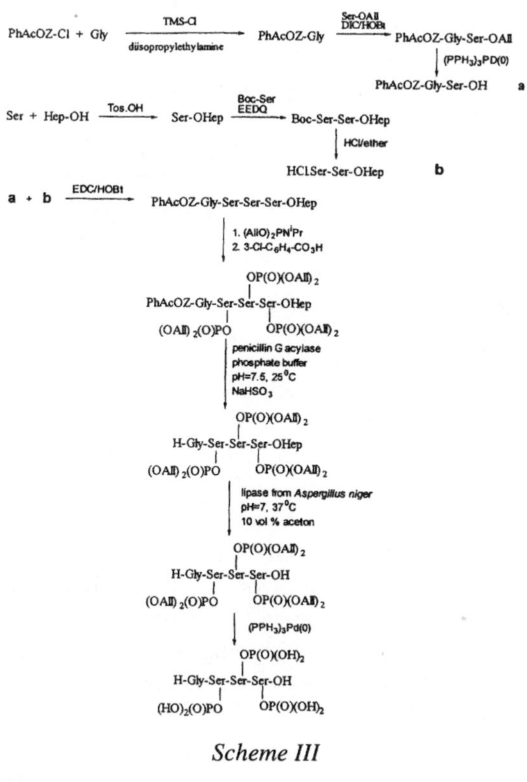

Scheme III

References

1. Hunter, T. and Martin, M., Cell, 70 (1992) 375.
2. Waldmann, H., Broun, P. and Kunz, H., Biomed. Biochim. Acta., 50 (1991) 243.

Synthesis of phosphorylated polypeptide by a thioester method

Saburo Aimoto, Koki Hasegawa, Xiang-Qun Li and Toru Kawakami
Institute for Protein Research, Osaka University, 3-2 Yamadaoka, Suita, Osaka, 565-0871, Japan

Introduction

Part of the ongoing research in this laboratory involves the development of a method for protein synthesis in which partially protected peptide thioesters are used as building blocks. The thioester method has been applied to the preparation of polypeptides with or without cysteine residues [1-3]. To make the method applicable to the preparation of phosphorylated polypeptides with molecular weights of around ten thousand, we investigated a procedure to prepare a partially protected phosphopeptide thioester and the coupling conditions under which the reaction proceeds at a reasonable rate without significant loss of the Acm groups on cysteine residues even in the presence of a silver compound.

Results and Discussion

We chose a phosphorylated partial sequence of the cAMP response element binding protein 1 (19-106), [Thr(PO_3H_2)^{69,71}]-CRE BP1(19-106)-NH_2 as a target peptide, as shown in Fig. 1 [4]. The [Thr(PO_3H_2)^{69,71}]-CRE BP1(19-106)-NH_2 was divided into three peptide segments for synthesis . The three partially protected peptide segments, Boc-[Lys(Boc)$^{23, 46, 48, 54}$, Cys(Acm)$^{27, 32}$]-CRE BP1(19–56)-SCH_2CH_2CO-β-Ala-NH_2 (**1**), Fmoc-[Thr(PO_3H_2)$^{69, 71}$, Lys(Boc)^{77}, Cys(Acm)^{76}]-CRE BP1(57–83)- SCH_2CH_2CO-β-Ala-NH_2 (**2**), [Lys(Boc)$^{97, 98, 105, 106}$]-CRE BP1(84–106)-NH_2 (**3**), were prepared using the Boc-based solid-phase method. Both of the C-terminal amino acid residues of peptides **1** and **2** are glycine residues to avoid racemization during segment condensation. Peptide **2** was prepared according to the procedure shown in Fig. 2. A fully protected peptide resin, which was obtained after completion of chain assemblage, was treated with TFMSA containing thioanisole, *m*-cresol, EDT and TFA at 0 °C for 3 h and at 20 °C for 1 h. A product, Fmoc-[Thr(PO_3H_2)$^{69, 71}$, Cys(Acm)^{76}]-CRE BP1(57–83)-SCH_2CH_2CO- β-Ala-NH_2, was isolated by reversed phase HPLC as shown Fig. 3 and freeze-dried to give a powder. A Boc group was introduced to the purified peptide. Peptide **2** thus obtained was isolated by a reversed phase HPLC in 2% yield, based on the C-terminal Gly residue. The cyclopentyl group in Thr($PO_3c$$Pen_2$) [5] and the cycloheptyl group in Asp(O*c*Hep) [6] were sufficiently stable under TFA treatment conditions during peptide chain elongation cycles and could be removed by TFMSA/TFA treatment [7]. However, the yield of this preparation was very low.

```
                                                        19
                                                        Met-Ser-   20
Asp-Asp-Lys-Pro-Phe-Leu-Cys-Thr-Ala-Pro-Gly-Cys-Gly-Gln-Arg-       35
Phe-Thr-Asn-Glu-Asp-His-Leu-Ala-Val-His-Lys-His-Lys-His-Glu-       50
Met-Thr-Leu-Lys-Phe-Gly-Pro-Ala-Arg-Asn-Asp-Ser-Val-Ile-Val-       65
                   P       P
Ala-Asp-Gln-Thr-Pro-Thr-Pro-Thr-Arg-Phe-Leu-Lys-Asn-Cys-Glu-       80
Glu-Val-Gly-Leu-Phe-Asn-Glu-Leu-Ala-Ser-Pro-Phe-Glu-Asn-Glu-       95
Phe-Lys-Lys-Ala-Ser-Glu-Asp-Asp-Ile-Lys-Lys-NH2       106
```

Fig. 1.The amino acid sequence of phosphorylated cAMP response element binding protein 1 (19-106)-NH_2, [Thr(PO_3H_2)^{69,71}]-CRE BP1(19-106)-NH_2. Thr(P) denotes phosphothreonine. Arrows indicate the sites of segment coupling.

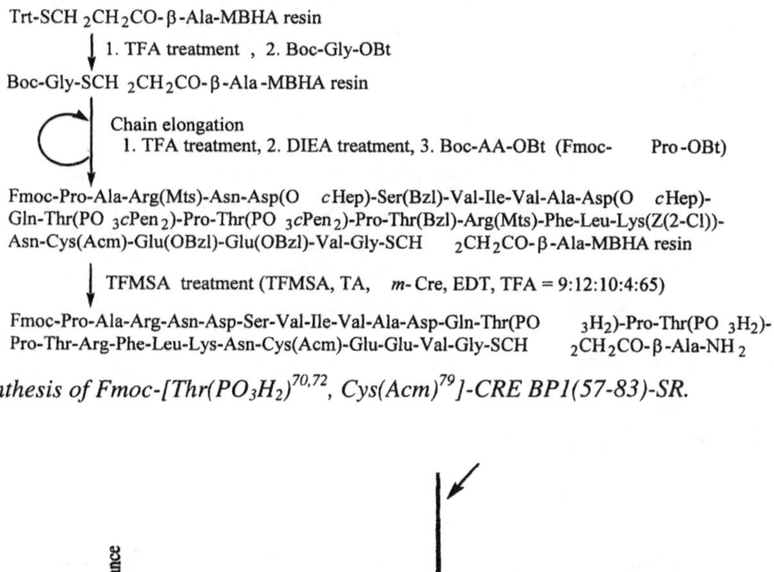

Fig. 2.Synthesis of Fmoc-[Thr(PO_3H_2)^{70,72}, Cys(Acm)^{79}]-CRE BP1(57-83)-SR.

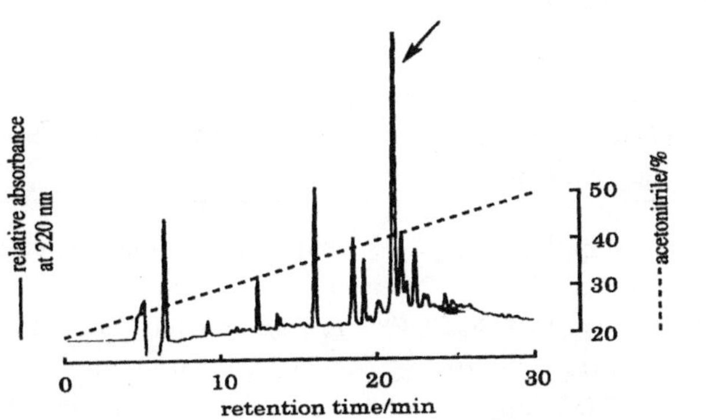

Fig. 3.HPLC elution profile of the crude product of Fmoc-[Cys(Acm)^{79}, Thr(PO_3H_2)^{69,71}]-CRE BP1(57-86)-SCH_2CH_2CO-β-Ala-NH_2. An arrow indicates the peak that contains a desired product. column: Cosmosil 5C_{18}AR (10 x 250 mm), eluent: 0.1% TFA in aq acetonitrile, 2.5 mL/min.

On the other hand, when the protected peptide resin was treated with anhydrous HF under dilute peptide conditions, the desired des-phosphate peptide, Fmoc-[Cys(Acm)^{76}]-CRE BP1(57–83)-SCH_2CH_2CO-β-Ala-NH_2, was obtained in good yield. This suggests that the peptide chain assembly proceeded without a serious problem, but some side reaction seemed to have occurred during the TFMSA/TFA treatment. Two main causes might be responsible for this low yield. One is the acid-resistant nature of the cycloheptyl group in Asp(O*c*Hep). It required long TFMSA/TFA treatment. The other is the modification of Fmoc group with benzyl and *o*-chlorobenzyl cations during TFMSA/TFA treatment of the protected peptide resin [8]. These side reactions were confirmed by mass analyses of the partially protected peptide and the isolated adduct after piperidine treatment of the peptide. The chemical structures and yields of building blocks used for the synthesis of [Thr(PO_3H_2)$^{69, 71}$]-CRE BP1(19-106)-NH_2 are summarized in Table 1. The yields were calculated based on the C-terminal amino acid residue in the resins or on the amino group in the starting MBHA resin.

Table 1 Building blocks for the synthesis of CRE BP1(19-106)-NH_2.

Building Blocks	Yield (%)
Boc-[Cys(Acm)^{27,32}, Lys(Boc)^{23,46,48,54}, Thr(PO_3H_2)^{69,71}]-CRE BP1(19-56)-SR (1) Boc-Met-Ser-Asp-Asp-Lys(Boc)-Pro-Phe-Leu-Cys(Acm)-Thr-Ala-Pro-Gly-Cys(Acm)-Gly-Gln-Arg-Phe-Thr-Asn-Glu-Asp-His-Leu-Ala-Val-His-Lys(Boc)-His-Lys(Boc)-His-Glu-Met-Thr-Leu-Lys(Boc)-Phe-Gly-SCH_2CH_2CO-β-Ala-NH_2	2%
Fmoc-[Cys(Acm)^{79}, Lys(Boc)^{77}, Thr(PO_3H_2)^{69,71}]-CRE BP1(57 3)-SR (**2**) Fmoc-Pro-Ala-Arg-Asn-Asp-Ser-Val-Ile-Val-Ala-Asp-Gln-Thr(PO_3H_2)-Pro-Thr(PO_3H_2)-Pro-Thr-Arg-Phe-Leu-Lys(Boc)-Asn-Cys(Acm)-Glu-Glu-Val-Gly-SCH_2CH_2CO-β-Ala-NH_2	2%
[Lys(Boc)^{97,97,105,106}]-CRE BP1(84 06)-NH_2 (**3**) Leu-Phe-Asn-Glu-Leu-Ala-Ser-Pro-Phe-Glu-Asn-Glu-Phe-Lys(Boc)-Lys(Boc)-Ala-Ser-Glu-Asp-Asp-Ile-Lys(Boc)-Lys(Boc)-NH_2	6%

Using these three building blocks, segment condensation was carried out according to the scheme shown in Fig. 4. Peptides **2** (0.44 μmol) and **3** (0.56 μmol) were condensed in the presence of AgCl (8.1 μmol), HOObt (19.6 μmol) and DIEA (2.2 μmol) in DMSO (67 μL) with stirring for 24 h at room temperature (fig. 5). To the reaction mixture, DTT (81 μmol) was added to quench the silver ion activity, and then piperidine (12 μL) was added to remove Fmoc groups. The product, [Cys(Acm)^{79}, Lys(Boc)$^{77, 97, 98, 105, 106}$, Thr(PO_3H_2)$^{69, 71}$]-CRE BP1(57-106)-NH_2, was isolated by reversed phase HPLC in 72% yield and used for the next coupling reaction after freeze-drying. This peptide (0.38 μmol), peptide **1** (0.64 μmol), AgCl (2.0 μmol), HOObt (21 μmol), and DIEA (14 μmol) were mixed in DMSO (68 μL) and the solution was stirred for 25 h (Fig. 6). After adding DTT to the reaction mixture, ether was added to precipitate a crude product. The crude product was treated with TFA containing 1,4-butanedithiol (5%) and the product, [Cys(Acm)$^{27, 32, 79}$, Thr(PO_3H_2)$^{69, 71}$]-CRE BP1(19-106)-NH_2, was isolated by reversed phase HPLC in 50% yield. This peptide (0.16 μmol), was treated with $AgNO_3$ (4.5 μmol),and DIEA (4.5 μmol) in a mixed solvent of water (60 μl) and trifluoroethanol (240 μl) for 5 h at room temperature. To the solution, DTT and then 1M HCl were added to decompose the silver thiolate. The final product was obtained after purification by reversed phase HPLC in the overall yield of 20%. Found *m/z* 10110.6 $(M+H)^+$ (average). Calcd *m/z* 10110.2 $(M+H)^+$ (average). Amino acid analysis: $Asp_{11.9}$ $Thr_{4.93}$ $Ser_{3.73}$ $Glu_{10.2}$ $Pro_{4.55}$ Gly_4 $Ala_{6.22}$ $1/2Cys_{2.15}$ $Val_{3.40}$ $Met_{1.75}$ $Ile_{1.63}$ $Leu_{6.32}$ $Phe_{6.89}$ $Lys_{8..45}$ $His_{3.63}$ $Arg_{3.03}$.

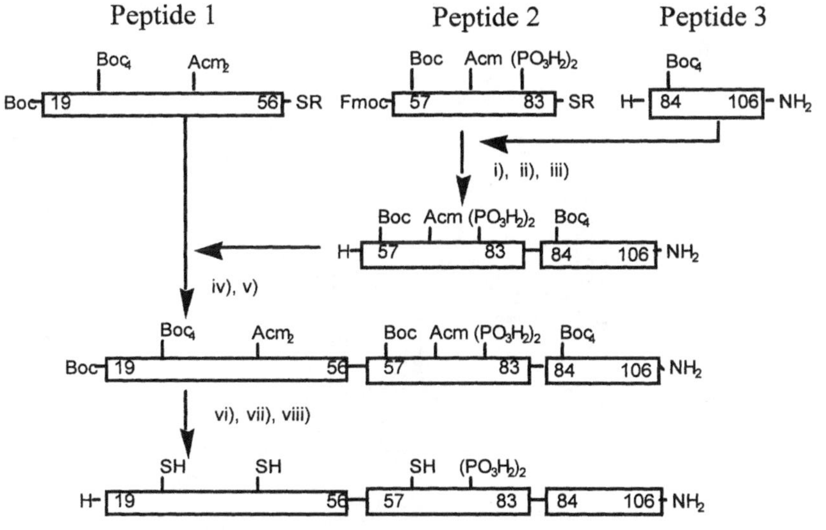

Fig. 4.A schematic showing the route for the preparation of [Thr(PO_3H_2)$^{69, 71}$]-CRE BP1(19-106)-NH_2. i) AgCl + HOObt + DIEA, ii) DTT, iii) piperidine, iv) AgCl + HOObt + DIEA, v) DTT, vi) 5% 1,4-butanedithiol / TFA, vii) $AgNO_3$ + DIEA + H_2O, viii) 1M HCl.

In this synthesis, we used silver chloride as an activator of the thioester moiety. Silver chloride is only slightly soluble in DMSO, the coupling reaction proceeded at a reasonable rate with no undesirable removal of Acm groups. Thus, silver chloride is a useful activating reagent of a thioester moiety to avoid the decomposition of Acm groups during segment condensation [9]. Before removal of the Fmoc group, DTT was added to avoid undesirable cleavage of Acm group caused by silver ions and piperidine as a nucleophile. At the final stage of this synthesis, the Acm group could be successfully removed in the presence of silver ions and water under neutral conditions [3], followed by 1M HCl treatment.

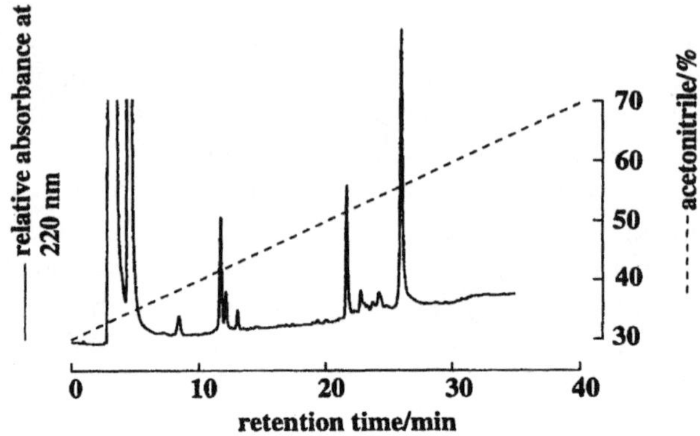

Fig. 5. The elution profile of the reaction mixture of Fmoc-[Cys(Acm)[79], Lys(Boc)[77, 97, 98, 105, 106], Thr(PO_3H_2)[69, 71]]-CRE BP1(57—106)-NH_2. An arrow indicates the peak containing a desired product. Column: Cosmosil 5C_{18}ARII (4.6 × 250 mm), eluent: 0.1% TFA in aq acetonitrile, 1.0 ml/min.

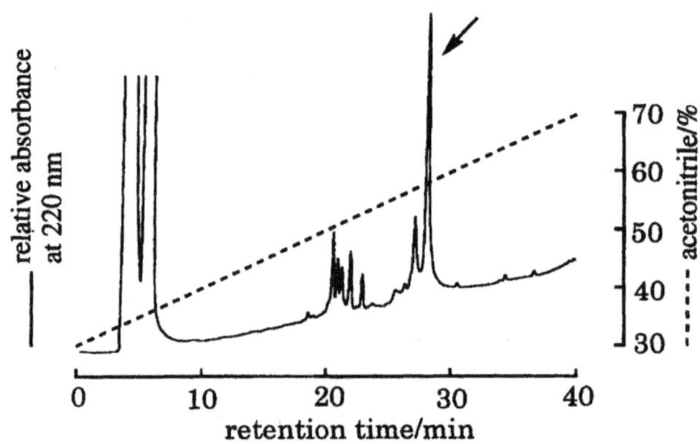

Fig. 6. The elution profile of the reaction mixture of Boc-[Cys(Acm)[27, 32, 79], Lys(Boc)[23, 46, 48, 54, 77, 97, 98, 105, 106], Thr(PO_3H_2)[69, 71]]-CRE BP1(19-106)-NH_2. An arrow indicates the peak containing a desired product. Column: Cosmosil 5C_{18}ARII (4.6 × 250 mm), eluent: 0.1% TFA in aq acetonitrile, 1.0 mL/min.

In conclusion, since we can now prepare a partially protected peptide thioester containing phosphate groups, a phosphorylated polypeptide can be obtained by the thioester method without problems. However, we have to improve the efficiency of the method for the preparation of partially protected peptide thioesters containing phosphate groups.

Acknowledgments

This research was partially supported by the grant-in-aid for Scientific Research on Priority Areas No. 06276102 from the Ministry of Education, Science, Sports and Culture.

References

1. Hojo, H. and Aimoto, S., Bull. Chem. Soc. Jpn., 64 (1991) 111.
2. Hojo, H., Yoshimura, S., Go, M. and Aimoto, S., Bull. Chem. Soc. Jpn., 68 (1995) 330.
3. Kawakami, T., Kogure, S. and Aimoto, S., Bull. Chem. Soc. Jpn., 69 (1996) 3331.
4. Maekawa, T., Sakura, H., Ishii, C.K., Sudo, T., Yoshimura, T., Fujisawa, J., Yoshida, M. and Ishii, S., EMBO J., 8 (1989) 2023.
5. Wakamiya, T., Togashi, R., Nishida, T., Saruta, K., Yasuoka, J.J., Kusumoto, J., Aimoto, S., Kumagaye, K.Y., Nakajima, K. and Nagata, K., Bioorg. Med. Chem., 5 (1997) 135.
6. Fujii, N., Nomizu, M., Futaki, S., Otaka, A., Funakoshi, S., Akaji, K., Watanabe, K. and Yajima, H., Chem. Pharm. Bull., 34 (1986) 864.
7. Fujii, N., Ohtaka, A., Ikemura, O., Hatano, M., Okamachi, A., Funakoshi, S., Sakurai, M., Shioiri, T. and Yajima, H., Chem. Fharm. Bull., 35 (1987) 3447.
8. Semchuk, P.D., Kondejewski, H., Daniels, L., Wilson, I. and Hodges, R., The abstruct of 15th Americam Peptide Symposium, ESCOM, Leiden, 1997, p.162.
9. Kawakami, T. and Aimoto, S., Chem. Lett., (1997) 1157.

Difference between N-phospho-L- and D-methionine in ester exchange reaction with uridine

Yuan Ma, Guo-Fang Shi, Yu-Ping Feng and Yu-Fen Zhao
Bioorganic Phosphorus Chemistry Laboratory, Department of Chemistry, Tsinghua University, Beijing, 100084, China

Introduction

Which came first, proteins or nucleic acids and how did they form a relationship in which they depend on each other for existence? Why did nature chose L-amino acid and D-ribose for the backbone of life? These have been major questions in origin of life studies [1]. In recent years, Zhao and coworkers have found that N-(O,O-dialkyl)phosphorylated amino acids(DAP-aa) [2], unlike common amino acids, can undergo some interesting biomimic reactions, such as self-catalysis to form self-assembly oligopeptides and ester-exchange reactions with the alcohol on the phosphoryl group. This suggests that DAP-aa might have played an important role in the prebiotic synthesis of proteins and nucleic acids. A hypothesis has been proposed by Zhao and Cao that DAP-aa might be regarded as the unique original seed for nucleic acids and proteins [3]. According to this hypothesis, oligopeptides, nucleotides and oligonucleotides should form simultaneously in the reaction of DAP-aa with nucleosides. Some experimental evidence has been found to support this hypothesis [4, 5]. In this paper we report ^{31}P NMR tracing experiments of N-phosphoryl-L- and D-methionine with uridine. It is very interesting that the reaction rate of L-methionine is faster than that of D-methionine.

Results and Discussion

DAP-Met was dissolved in $CHCl_3$ at a concentration of 3.37 mol/L, and then 0.2 ml of the solution was added to an NMR tube. Uridine was dissolved in anhydrous pyridine at a concentration of 2 M, and 0.18 ml was transferred into the tube. About 0.001 g $(CH_3COCHCOCH_3)_3Cr$ was added as relaxation agent. The relative concentrations of each substrate and products were determined by integration of ^{31}P NMR spectra.

In Fig. 1, the ^{31}P NMR chemical shifts at 10 ppm is DAP-Met, peaks in 2-3 ppm are hydrolysis products, and –9 ppm is pyrophosphate, while 20 ppm represents five membered tetracoordinate phosphorus compound, as reference [6]. Table 1 shows the reaction rate constants (k) of DAP-Met with uridine. As we expected, it indicated that changing the alkyl group on phosphorus from propyl to isopropyl, decreased rate constants markedly. On the other hand, the reaction rate of L-Met was 4.5 times faster than that of D-Met. This result gives some indication that with the involvement of phosphorus, the D-ribose may tend to select L-amino acid.

Table 1. Reaction rate constants (k) of DAP-Met with uridine.

Compounds	$K \times 10^{-3}$ (h^{-1})
1 DIPP-L-Met	0.1
2 DPP-L-Met	37.2
3 DPP-D-Met	8.3

DIPP: diisopropylphosphoryl DPP: dipropylphosphoryl

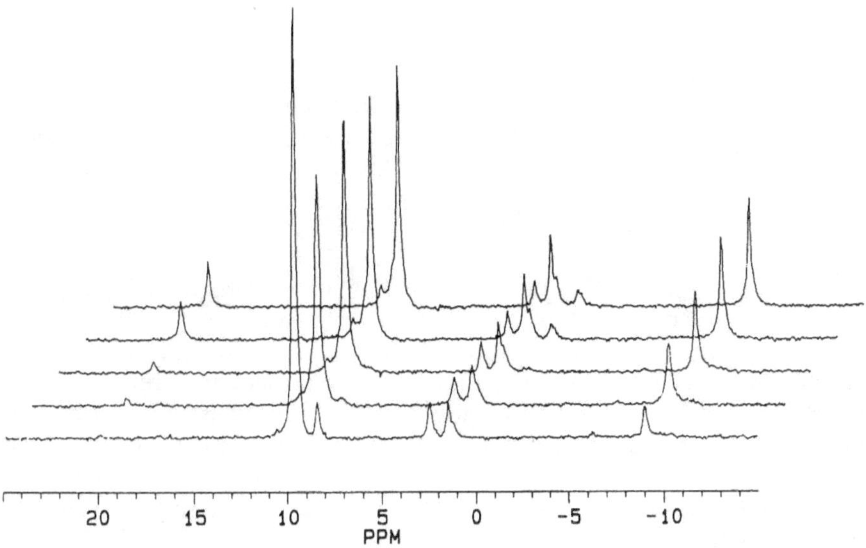

Fig. 1. ^{31}P NMR stack spectra for reaction of DPP-D-Met with uridine in anhydrous pyridine.

Acknowledgments

This work was supported by National Natural Science Foundation of China and Tsinghua University

References

1. Eigen, M., Naturwiss, 58 (1971) 465.
2. Zhao, Y.F., Ju, Y., Li, Y.M., Wang, Q., Yin, Y.W. and Tan, B., Int. J. Peptide Protein Res., 45 (1995) 514.
3. Zhao, Y.F. and Cao, P.S., J. Biol. Phys., 20 (1994) 283.
4. Zhou, W.H., Ju, Y., Zhao, Y.F., Wang, Q.G. and Luo, G.A., Origins of Life, 26 (1996) 547.
5. Ma, Y., Li, X.H., Chen, Y., Feng, Y.P. and Zhao, Y.F., Chin. Chem. Letts., 7 (1996) 905.
6. Zhao, Y.F., Yan, Q.J. and Wang, Q., Int. J. Peptide Protein Res., 47 (1996) 276.

Synthesis of analogs of phosphoamino acids and their biomimic reactions

Yong Ju, Dong-Yan Qin and Yu-Fen Zhao
Bioorganic Phosphorus Chemistry Laboratory, Department of Chemistry, Tsinghua University, Beijing, 100084, China

Introduction

Phosphorus, carbohydrate and protein play very important roles in biological chemistry. In order to understand their intrinsic relationships and their reaction mechanisms, many phosphoamino acids were synthesized and their chemical properties investigated [1-4] were found. Many interesting bioorganic characteristic reactions were found. In order to further confirm the significance of phosphorus in biochemistry, a series of model compounds were synthesized, including phosphoryl glucoaminic acids, analogs of phosphoryl amino acids containing P-C band. The recitatives with alcohol and amino acid were also investigated by NMR and MS methods. The results indicated that the reaction characteristics of phosphoryl organic small molecules depended on configuration, functional groups and positions.

Results and Discussion

Synthesis

(O, O-biphenyl phosphinyl-methylene)-butanedioic anhydride (**3**): Equal moles of itaconic acid and triphenyl phosphite were added to a round flask and then heated to 140 °C for 30 min. After cooling, the reaction mixture was recrystallized in petrol-ethyl acetate (1:1) to give colorless crystals (yield 75%).

N- (O, O-diisopropyl) phosphoryl glucoaminic acid (**6**): To a stirred aqueous solution of glucoamino acid, a little ethanol and triethyl amine were added to adjust the pH to11 and then diisopropyl phosphorochloridate in CCl_4 was added drop wise at 0 °C. After stirring for 24 hr, the reaction mixture was extracted with ethyl acetate, the residue aqueous layer was adjusted with diluted HCl to pH=4 in ice-salt bath. Then the aqueous layer was washed with CH_2Cl_2 and finally the residue solution was dried to give N-(O, O-diisopropylphosphoryl) glucoaminic acid (**6**).

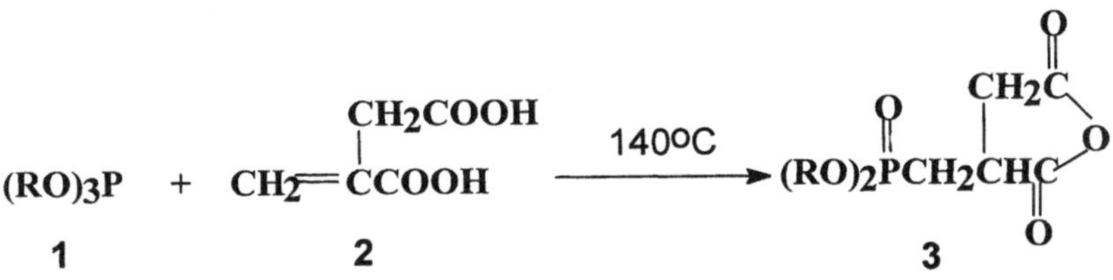

Scheme 1.

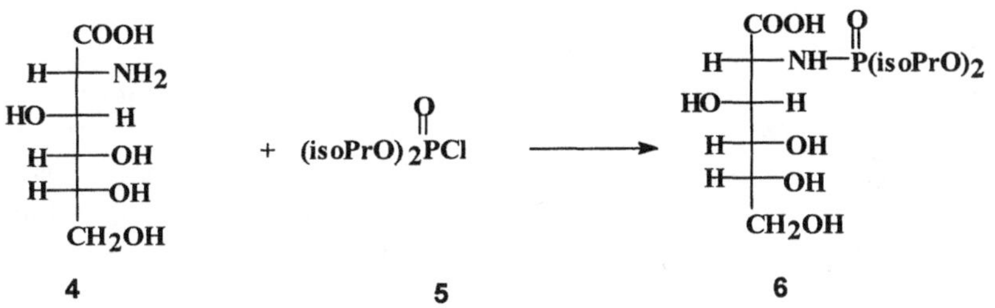

Scheme 2.

Structural Determination

3 mp. 107-108 °C. Molecular ion peak was indicated at m/z 346 [M^+] in the EI-mass spectrum. Element analysis: Found C 58.93, 58.84, H 4.44, 4.49% (Calculated for $C_{17}H_{15}O_6P$ C 58.93 H 4.34%). In its IR spectrum, there were anhydride absorption peaks at ν^{KBr} 1843 and 1787 cm^{-1} but no absorption of hydroxyl group. The ^{1}HNMR data are given in table 1.

*Table 1. ^1H NMR spectral data (chemical shifts δ ppm) of Compound **3**.*

	^{1}HNMR(coupling constants)
CH	3.59 (m,1H)
CH_2CO	3.31 (dd,1H,Jgem=-19.0,J=9.8Hz)
	3.10 (dd,1H,Jgem=-19.0,J=7.9Hz)
CH_2PO	2.83 (ddd,1H,$^2J_{P\text{-}H}$=19.0,Jgem=-15.6,J=3.0Hz)
	2.38 (ddd,1H,$^2J_{P\text{-}H}$=17.0,Jgem=-15.6,J=10.9Hz)
ArH	7.12 (m,10H)

There is one signal peak with $^2J_{P\text{-}H}$ 19.0 and 17.0 Hz in ^{31}PNMR spectrum of **3**. From the above spectrum data, structure of **3** was established as (O, O-diphenyl phosphinyl methylene) butanedioic anhydride (fig. 1). ^{13}CNMR data are shown in Fig.1. To our knowledge, it is a new compound.

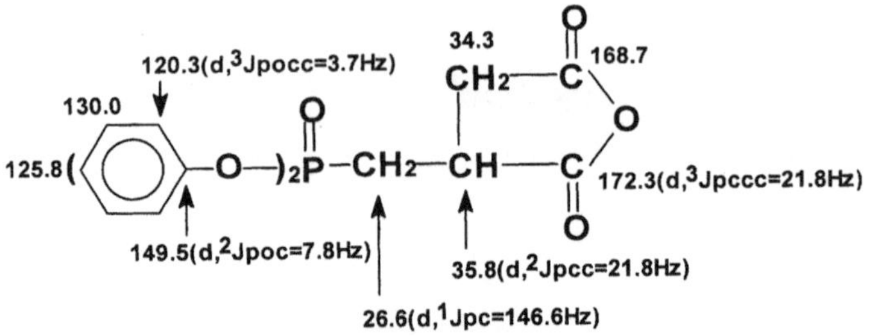

Fig. 1.The structure of ***3*** *and its* ^{13}*CNMR spectral data*
(Coupling constants, Hz).

Compound **6**: The high-resolution negative ion FAB mass spectrum (HRMS) gave $[M\text{-}H]^-$ at m/z 358.1264 indicating the molecule formula of **6** to be $C_{12}H_{29}O_9NP$. The FAB MS gave quasimolecular ion peak at m/z: 358 $[M\text{-}H]^-$ and corresponding fragments ion peaks $[M\text{-}i\text{-}PrO\text{-}H]^-$ and $[M\text{-}2Xi\text{-}PrO\text{-}H]^-$. The ^{31}PNMR spectrum (decoupling) gave a signal at δ 7.36 ppm. The above spectral data were identical with the target molecule.

Chemical Properties

Exchange reaction with n-BuOH: Compound **3** and **6** were added to n-BuOH and kept for a day as specialized in the literature [6], and then the reaction mixtures were analyzed for the exchange products by ^{31}PNMR spectrum and FAB-MS. The result showed that exchange products were only found in the reaction of compound **6**.

Reaction with amino acid: **3** was reacted with amino acids or amino acid esters under the same condition [2, 4], the reaction mixture was analyzed by FAB mass spectrum and indicated no peptide formation.

Conclusion

The investigation showed that the chemical properties of analogs of phosphoamino acid with P-C, P-O, and P-N bands are very different. The results indicated that the reaction characteristics of phosphoryl biological small molecules not only depended on configuration, functional groups and positions, but also on the type of atoms. Further investigations are in progress.

Acknowledgements

The authors thank the National Natural Science Foundation of China and Tsinghua University for the financial support.

References

1. Xue, C.B. and Zhao, Y.F., J. Chem. Soc., Perkin Trans., 2 (1990) 431.
2. Zhao, Y.F., Li, Y.M., Yin, Y.W. and Li, Y.C., Science in China, 36 (1993) 1451.
3. Ju, Y., Zhao, Y.F., Sha, Y.W. and Tan, B., Phosphorus, Sulfur and Silicon, 101 (1995) 117.
4. Zhao, Y.F., Ju, Y., Li, Y.M., Yin, Y.W., Wang, Q. and Tan, B., Int. J. Pept. Protein Res., 45 (1996) 456.
5. Henry, B.F. and Michael, J.S., Biochem. J., 141 (1974) 715.
6. Tan, B. and Zhao, Y.F., Youji Huaxue, 15 (1995) 30.

Solid phase synthesis of heterocyclic combinatorial libraries derived from peptides

John M. Ostresh, Christa C. Schoner, Vince T. Hamashin, Marc Giulianotti, Adel Nefzi, Michael J. Kurth and Richard A. Houghten

Torrey Pines Institute for Molecular Studies, 3550 General Atomics Ct., San Diego, CA 92121, USA

Introduction

Since the introduction of the simultaneous multiple peptide synthesis method ("teabag synthesis") [1], our laboratory has not only introduced, but also extensively developed, many of the key techniques used in the combinatorial field today. Following the introduction of synthetic peptide combinatorial libraries by this [2] and other laboratories [3], we later demonstrated practical means for deconvolution of large soluble mixtures to identify active compounds using both the iterative [2] and positional scanning formats [4]. Mixture-based synthetic combinatorial libraries have been prepared primarily by the resin-mixing approach (referred to as the "divide, couple and recombine" or "split-synthesis" method) [2, 3, 5] and the reagent mixture approach [4, 6]. In addition, our laboratory has extensively developed the "libraries from libraries" concept introduced in 1994 [7, 8]. Using this concept, the diversity of existing peptide combinatorial libraries can be leveraged to generate additional combinatorial libraries having completely different physical and chemical characteristics. This concept allows one to leverage not only the diversity of existing libraries but also the extensive effort that has gone into the development of these existing libraries. We have extended this approach for the generation of large mixture-based acyclic and heterocyclic combinatorial libraries starting from existing peptide libraries [9]. The techniques currently in use in our laboratory for the preparation of synthetic combinatorial libraries derived from peptides (fig. 1) are described. Specifically, the general technique used to determine the ratios of reagents necessary for the generation of equimolar mixtures will be examined.

Results and Discussion

The positional scanning method of deconvolution offers excellent time and labor savings over the iterative deconvolution method. In the positional scanning method, sublibraries containing the same compounds are synthesized. The number of sublibraries corresponds to the number of positions of diversity. Each sublibrary differs in the manner in which the mixtures are pooled based on one substituent within the mixture being defined. Following screening of each sublibrary, the substituents from the most active mixtures are used to generate individual compounds. Individual active compounds can be identified within one to two weeks of the initial screening of the library. However, incorporation of defined positions with the library is prohibitively time consuming unless reagent mixtures are used for mixture positions following incorporation of the defined positions. This necessitates determination of an isokinetic ratio of the reagents, which is the ratio of the reagents necessary for equimolar incorporation of the required building blocks.

To illustrate the method used to determine isokinetic ratios, the competitive acylation of a resin-bound amine is used. A large excess of the acylating agents is necessary such that pseudo first order reaction kinetics is assumed. In order to achieve equimolarity of the reaction products, the reaction rates must be equal. In addition, the concentrations of acylating agents are constant while the concentration of the resin-bound amine is equal in both reactions since it is a competitive reaction. Thus,

$$[COOH]_b/[COOH]_a = k_a/k_b \qquad (1).$$

Therefore, the concentration of the two carboxylic acids necessary for equimolar incorporation is inversely proportional to the rate constants.

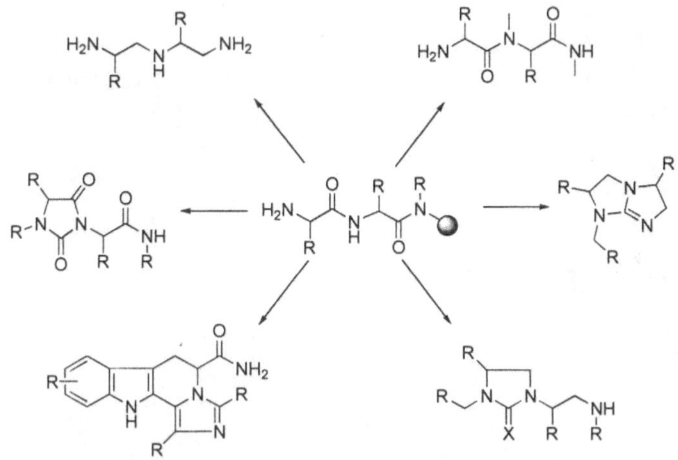

Fig. 1. General scheme demonstrating the "libraries from libraries" concept for generating both acyclic and heterocyclic combinatorial libraries from peptide library templates.

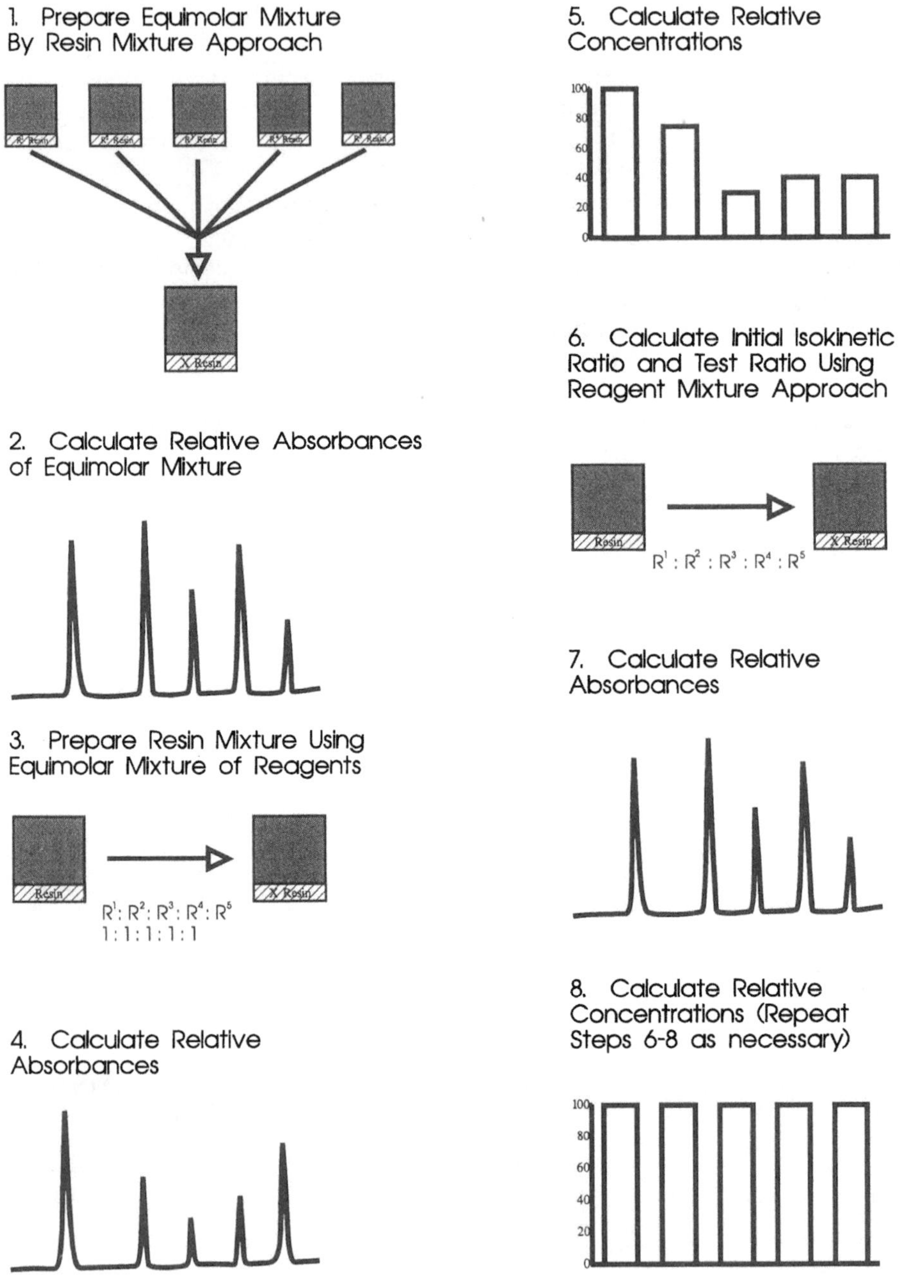

Fig. 2. General approach for determining isokinetic ratios used in reagent mixtures.

Table 1. Initial isokinetic ratios of aromatic aldehydes necessary for equimolar reductive alkylation of Ala-Phe-Lys(ClZ)-methylbenzhydrylamine resin and percent incorporation relative to 4-fluorobenzaldehyde. The reaction products are listed in the order of elution in Fig. 3.

No.	Aldehyde	Isokinetic Ratio	Percent Incorporation
1	4-quinolinecarboxaldehyde	1.09	128.8
2	3-hydroxybenzaldehyde	0.56	87.2
3	3-cyanobenzaldehyde	0.56	93.0
4	4-fluorobenzaldehyde	1.00	100.0
5	2-chloro-5-nitrobenzaldehyde	1.15	85.2
6	3-bromobenzaldehyde	0.50	90.8
7	3-trifluoromethylbenzaldehyde	0.68	104.7
8	3-nitro-4-chlorobenzaldehyde	0.63	112.7
9	3,4-dichlorobenzaldehyde	0.96	158.1
10	4-isopropylbenzaldehyde	1.08	89.4
11	4-phenoxybenzaldehyde	1.35	86.3

Once this relationship is established, the isokinetic ratios necessary for equimolar incorporation can be determined by studying the reaction when the carboxylic acids are present in equimolar amounts. In this case,

$$k_a/k_b=\ln([\text{product}]_a/[\text{product}]_b) \qquad (2).$$

Using Eqns. 1 and 2, the isokinetic ratios can then be determined. Errors in the determined ratios are exaggerated due to the inverse of the logarithmic function in the calculation. We have found an iterative approach involving synthesis and estimation of the isokinetic ratios to be more straightforward. Fig. 2 shows the generalized method for determining isokinetic ratios. An equimolar mixture is first made to determine the relative absorbance of the desired products under standard RP-HPLC conditions. A mixture resin is then synthesized using an equimolar reagent mixture and the relative concentrations of the products determined using the equimolar standard previously synthesized. Isokinetic ratios are then estimated. Finally, the isokinetic ratios are tested and the isokinetic ratios adjusted as necessary.

Table 1 shows the initial isokinetic ratios calculated for approximately equimolar reductive alkylation of Ala-Phe-Lys(ClZ)-methylbenzhydrylamine resin using aromatic aldehydes. The calculations were based on competitive reactions between each individual aldehyde and 4-fluorobenzaldehyde, which was used as the standard. Fig. 3 shows the RP-HPLC of the eleven products using the isokinetic ratios for the reaction described in Table 1. In addition, the calculated percent incorporation of each compound is listed. Under the conditions used, the assumption was made that reaction kinetics observed were pseudo first order for imine formation reaction. Based on the percent incorporation, only two benzaldehydes had concentrations that were more than 15% different from the 4-fluorobenzaldehyde. Since in many assays, orders of magnitude differences in activity are found for different mixtures, the selection criteria are clear. Therefore, absolute equimolarity is not necessary. In addition, ratios such as these would not cause wide variation from equimolarity if used in only one position of diversity.

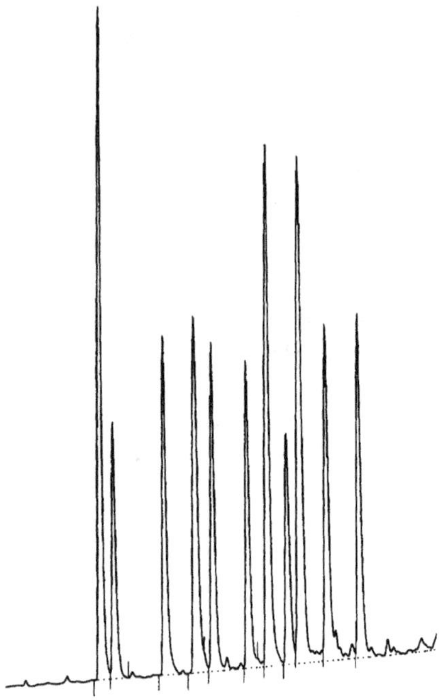

Fig. 3.RP-HPLC of the reductive alkylation products listed in Table 1.

Conclusion

We have found that the general method described is useful for calculating isokinetic ratios for relatively equimolar incorporation of a variety of reagents. We have calculated the ratios for over 40 benzaldehydes for the reductive alkylation of amines. In addition, we regularly use isokinetic ratios for the acylation of amines using more than 80 Boc amino acids, 60 Fmoc amino acids, and 70 carboxylic acids.

Acknowledgements

This work was supported in part by National Science Foundation grant CHE-9520142 (R.A. Houghten) and by Trega Biosciences, Inc., San Diego, California.

References

1. Houghten, R.A., Proc. Natl. Acad. Sci. U.S.A., 82 (1985) 5131.
2. Houghten, R.A., Pinilla, C., Blondelle, S.E., Appel, J.R., Dooley, C.T. and Cuervo, J.H., Nature, 354 (1991) 84.
3. Lam, K.S., Salmon, S.E., Hersh, E.M., Hruby, V.J., Kazmierski, W.M. and Knapp, R.J., Nature, 354 (1991) 82.

4. Pinilla, C., Appel, J.R., Blanc, P. and Houghten, R.A., Biotechniques, 13 (1992) 901.
5. Furka, A., Sebestyen, F., Asgedom, M. and Dibo, G., Int. J. Pept. Protein Res., 37 (1991) 487.
6. Ostresh, J.M., Winkle, J.H., Hamashin, V.T. and Houghten, R.A., Biopolymers, 34 (1994) 1681.
7. Ostresh, J.M., Husar, G.M., Blondelle, S.E., Dörner, B., Weber, P.A. and Houghten, R.A., Proc. Natl. Acad. Sci. USA, 91 (1994) 11138.
8. Cuervo, J.H., Weitl, F., Ostresh, J.M., Hamashin, V.T., Hannah, A.L. and Houghten, R.A., In Maia, H.L.S., (Ed.) Peptides 1994 (Proceedings of the Twenty-Third European Peptide Symposium), ESCOM, Leiden, The Netherlands, 1995, p. 465.
9. Nefzi, A., Ostresh, J.M., Meyer, J.P. and Houghten, R.A., Tetrahedron Letters, 38 (1997) 93.

Parallel synthesis of muramyl peptides derivatives

Suo-De Zhang, Gang Liu, Su-Quan Xia and Zhen-Kai Ding
Institute of Pharmacology and Toxicology, Academy of Military Medical Science, Beijing, 100850, China

Introduction

N-acetyl muramyl-L-alanyl-D-isoglutamine (MDP) is the minimal structure for the immunoadjuvancy elicited by Freund's Complete Adjuvant [1]. Although MDP has wide biological activities, its application is limited due to its short duration of action and undesired side effects [2]. In our attempt to study structure-activity relationships as well as to dissociate between immunomodulatery and other unwanted biological activities, we prepared 46 MDP derivatives in which organic acids were separately linked to the C terminal of MDP or to the C terminal of MDP's isomers: N-acetyl isomuramyl-L-alanyl-D-iso-glutamine through a bridge. As the generation of molecular diversity through combinatorial synthesis promises to be more efficient [3, 4, 5], and the solid phase approach offers a number of advantages over solution phase methods. We constructed the small library of MDP derivatives on solid phase with multipin kits [6]. Screening for biological activities of these compounds is in progress.

Results and discussion

Fmoc-D-isoglutamine was synthesized in liquid phase via the reaction of Fmoc-OSu with D-isoglutamine in aqueous solution and purified by silica column chromatography, m.p.204-205 °C. Fmoc-L-isoglutamine was also synthesized in the same way, m. p. 204-205 °C.

Our design intercalated lysine between MDP and organic acids as a spacer arm. Fmoc-Lys(Tfa)-OH was selected at first because the N_ε-trifluoacetyl group is unaffected by normal acidic deprotection conditions and by anhydrous piperidine ,but is readily cleaved by mild alkaline treatment using aqueous piperidine [7]. Its application on pin in our laboratory was not satisfactory as the N_ε-trifluoacetyl group could not be cleaved by aqueous piperidine. Fmoc-Lys(Dde)-OH was then selected for the reason that the Dde protecting group is essentially stable to acid and base conditions, but can be easily cleaved by 2% hydrazine .

MDP derivatives were synthesized according to the following route.

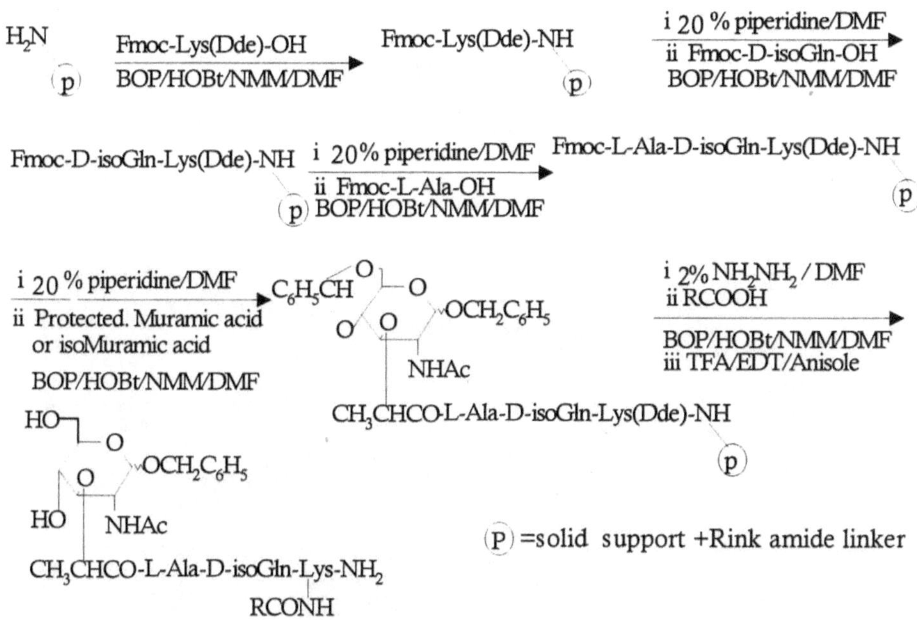

Fig. 1. Synthetic routes of MDP derivatives.

Table1. MDP derivatives in which protected muramic acid was used.

Entry	RCOOH	MW calc. (FAB.MS)	Purity	Entry	RCOOH	MW calc. (FAB.MS)	Purity
B1	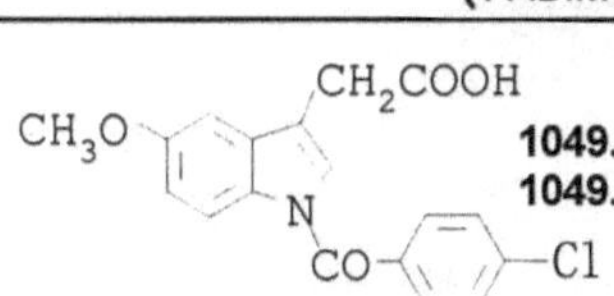	**1049.7** **1049.6**	c	**B2**	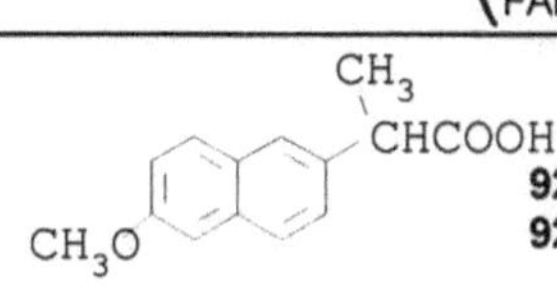	**922.2** **922.4**	b
B3	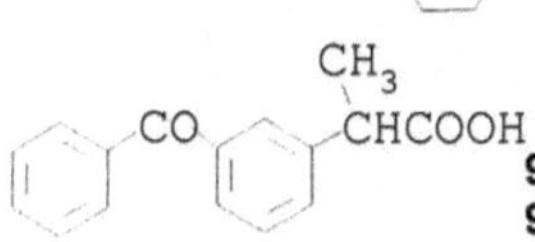	**946.2** **946.4**	c	**B4**	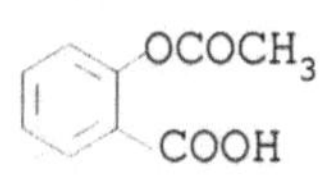	**872.1** **872.3**	d
B5	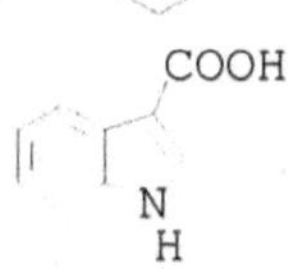	**853.1** **853.3**	b	**B6**	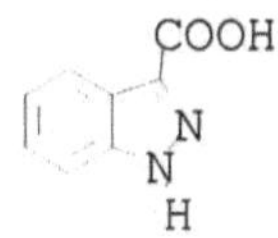	**854.1** **854.3**	b
B7	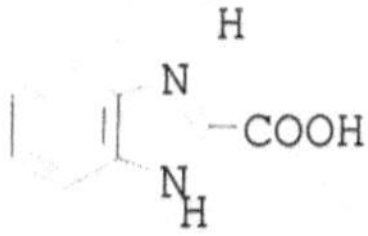	**854.1** **854.2**	a	**B8**	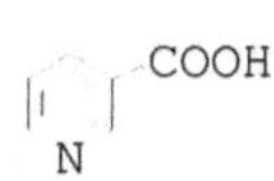	**815.0** **815.2**	a

Table1. (continued).

Entry	RCOOH	MW (calc. / FAB.MS)	Purity	Entry	RCOOH	MW (calc. / FAB.MS)	Purity
B9	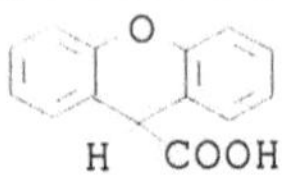	**917.9** **918.3**	b	**B10**	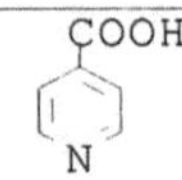	**815.0** **815.3**	b
B11	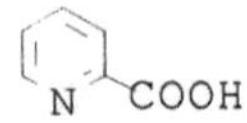	**815.0** **815.2**	a	**B12**	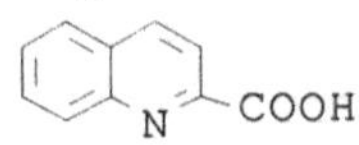	**865.1** **865.4**	b
C1	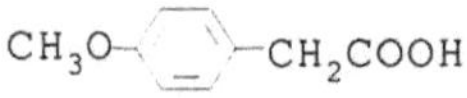	**858.1** **858.2**	a	**C2**	$CH_3CH{=}CHCH{=}CHCOOH$	**804.1** **804.3**	b
C3	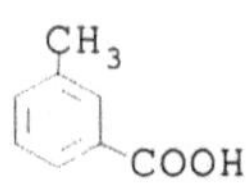	**828.0** **828.2**	b	**C4**	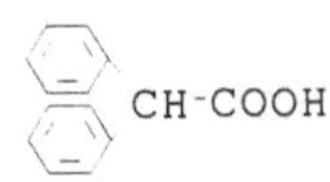	**904.2** **904.3**	b
C5	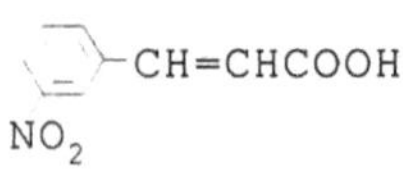	**885.1** **885.0**	b	**C6**	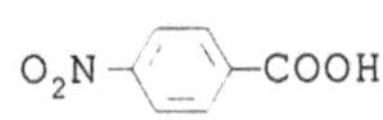	**859.0** **859.2**	a
C7	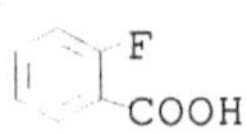	**831.9** **832.3**	a	**C8**	$CH_3-CH(OH)-CH_2-COOH$	**796.0** **796.0**	b
C9	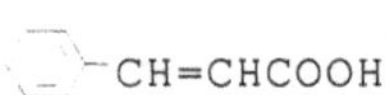	**840.1** **840.2**	a	**C10**	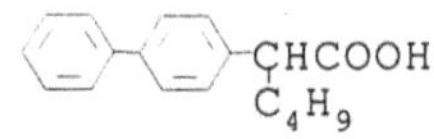	**959.9** **960.3**	c
C11	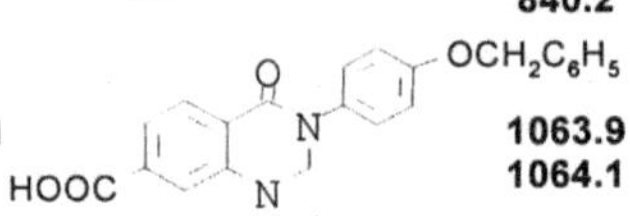	**1063.9** **1064.1**	c	**C12**	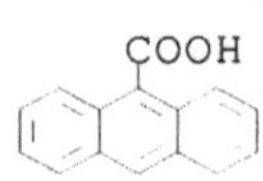	**914.2** **914.0**	c
D1	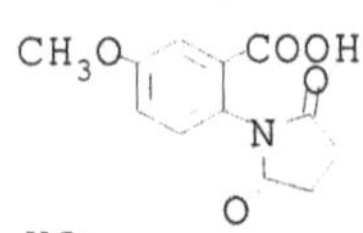	**940.9** **941.3**	c	**D2**	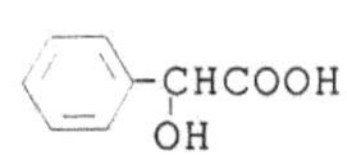	**844.1** **844.3**	a
D3	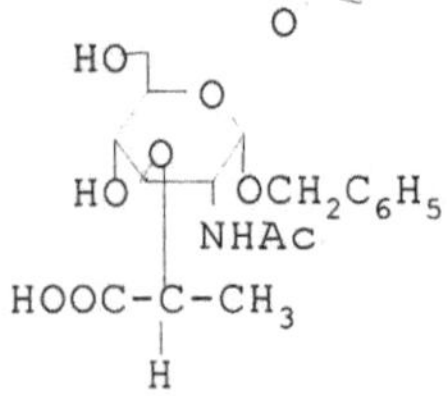	**1075.4** **1075.3**	b	**D4**	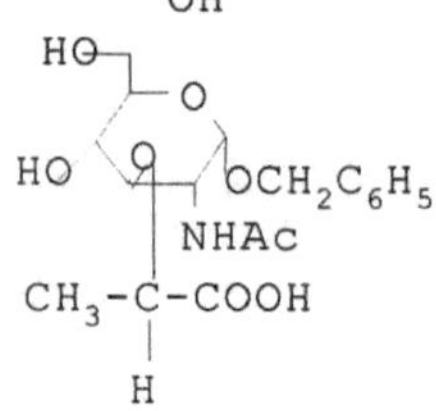	**1075.4** **1075.3**	b
D5	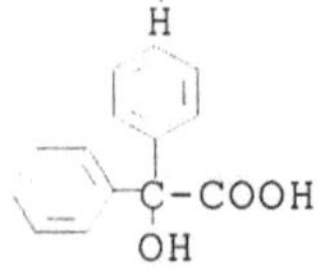	**920.2** **920.4**	b	**D6**	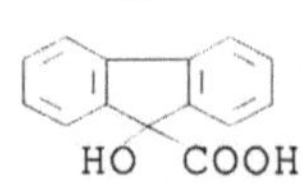	**917.9** **918.4**	b
D7	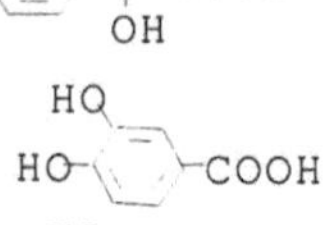	**862.1** **862.2**	c	**D8**	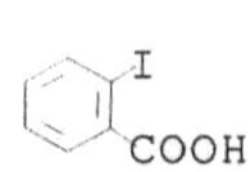	**939.9** **940.2**	b

Purity was determined by HPLC. Key: a=purity>90%, b=purity 80-90%, c=purity 70-80%, d=purity<70%

Table 2. MDP derivatives in which protected isomuramic acid was used.

Entry	RCOOH	MW calc. (FAB.MS)	Purity	Entry	RCOOH	MW calc. (FAB.MS)	Purity
E1	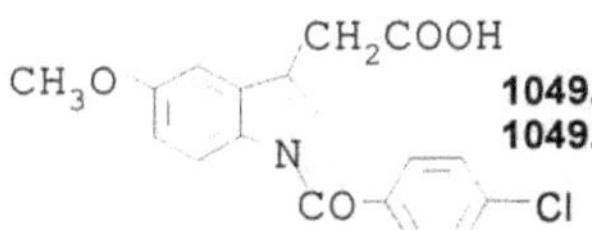	1049.7 1049.6	b	E2	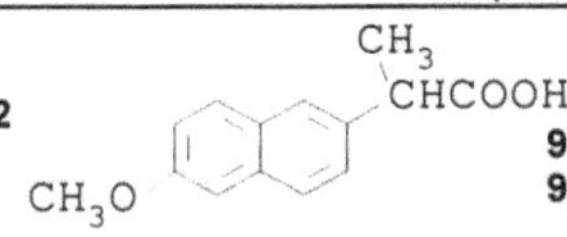	922.2 922.4	b
E3	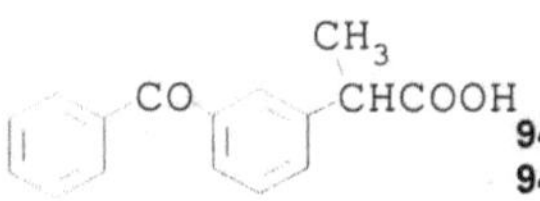	946.2 946.4	c	E4	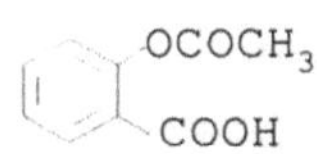	872.1 872.3	d
E5	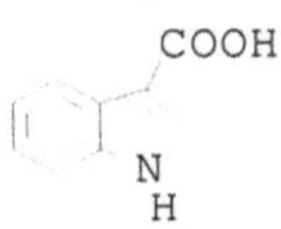	853.1 853.3	b	E6	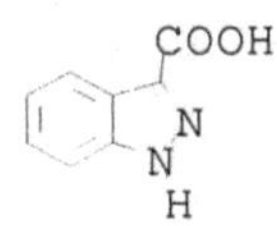	854.1 854.3	c
E7	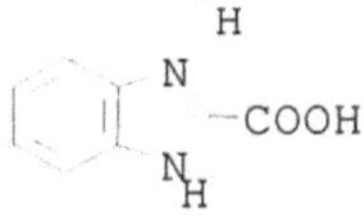	854.1 854.2	a	E8	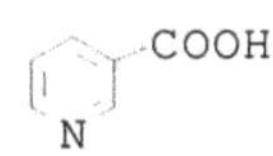	815.0 815.2	a
E9	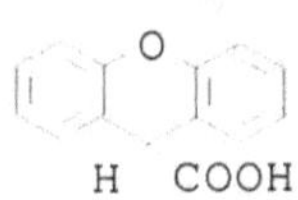	917.9 918.3	b	E10	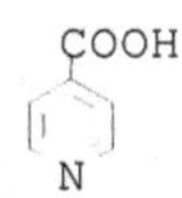	815.0 815.3	b
E11		815.0 815.2	b	E12		865.1 865.4	b
F1	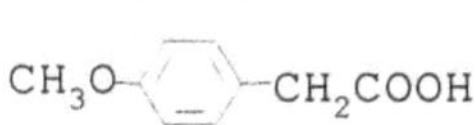	858.1 858.2	a	F2	$CH_3CH{=}CHCH{=}CHCOOH$	804.1 804.3	b

1. Ellouz, F., Adam, A. and Ciorbaru, R., Biochem. Biophys. Res. Commun., 59 (1974) 1317.
2. Baschang, G., Tetrahedron, 45 (1989) 6331.
3. Terrett, N.K., Gardner, M. and Gordon, D. W., Tetrahedron, 51 (1995) 8135.
4. Nefzi, A., Ostresh, J.M. and Houghten, R.A., Chem. Rev., 97 (1997) 449.
5. Thompson, L.A. and Ellman, J.A., Chem. Rev., 96 (1996) 555.
6. Geysen, H.M., Meloen, R.H. and Barteling, S.J., Proc. Natl. Acad. Sci. USA, 81 (1984) 3998.
7. Atherton, E. and Sheppard, R.C., Solid phase synthesis: a practical approach, 1989, p. 53.
8. Bycroft, B.W., Chan, W.C. and Chhabra, S.R., J. Am. Chem. Soc., (1993) 778.

β-Cyclodextrin for presentation of bioactive peptides to molecular recognition

Norbert Schaschke [a], Stella Fiori [a], Hans-Jürgen Musiol [a], Irmgard Assfalg-Machleidt [b], Werner Machleidt [b], Chantal Escrieut [c], Daniel Fourmy [c], Gerhard Müller [d] and Luis Moroder [a]

[a]Max-Planck-Institut für Biochemie, D-82152 Martinsried; [b]Institut für Physiologische Chemie, Physikalische Biochemie und Zellbiologie, LMU München, D-80336 München; [c]INSERM U152, CHU Rangueil, F-31054 Toulouse; [d]Bayer AG, MD-IM-FA, Q18, D-51368 Leverkusen

Introduction

CCK-B/gastrin receptors are found throughout the central nervous system where they regulate anxiety/panic attacks and dopamine release implicated in the pathogenesis of dopaminergic related behavioral disorders in humans. In the periphery these receptors regulate acid and histamine secretion as well as growth in the gastric mucosa and gastrointesinal motility [1]. Because of the multiple physiological and pathophysiological implications, the CCK-B/gastrin receptor represents an attractive target for drug development. However, to rationally design highly selective agonists and antagonists for human medicine, ligand binding site(s) have to be identified and precisely delineated. The CCK-B receptor belongs to the large family of G-protein-coupled receptors (GPCR) the topology of which consists of a bundle of seven transmembrane helices tethered by a series of extracellular and cytoplasmatic loops of variable lengths [2]. The footprint of ligand binding derived from mutagenesis studies is spatially rather conserved and involves residues of the extracellular loops as well as residues located in more hydrophobic compartments of the transmembrane domain [2-4].

By using lipophilic gastrin derivatives that contain di-acylatty-glycerol moieties as membrane anchor irreversible trapping of gastrin on the cell membrane has been achieved. Consequently, recognition by and binding to the receptor has been restricted to a membrane-bound mechanism, i.e. to the two-dimensional diffusion of the ligand on the membrane surface and lateral docking to the receptor binding site(s) [5-8]. Using as experimental constraints the resulting receptor binding affinities as well as information gained by mutagenesis studies of the CCK-B receptor, docking of the hormone gastrin to the homology modeling-derived structure of the human CCK-B/gastrin receptor [9] gave a picture of the ligand binding mode that shows a large portion of the peptide spanning the extracellular surface of the receptor, with the C-terminal tetrapeptide amide of the ligand, i.e. the message part of gastrin [10], inserted into the helix bundle [8, 11].

Conformational studies on lipogastrin anchored on artificial lipid bilayers exclude a membrane-induced prefolding in the receptor recognition pathway for this peptide hormone as expected from a thermodynamic point of view [11]. A direct collisional event from the extracellular space is suggested to be responsible for the receptor-mediated signal transduction. In the present study we made use of cyclodextrin/gastrin conjugates to deliberately impede preadsorption of the ligand on the cell membrane and to restrict via the hydrophilic carbohydrate moiety the receptor recognition pathway to a collisional event from the extracellular space .The penetration of the receptor helix bundle was modulated by a bulky template [12]. From receptor affinity data additional experimental facts were expected to delineate and validate ligand binding site(s) on the receptor model.

Results and Discussion

Covalent linkage of bioactive peptides to cyclodextrins has been proposed [13] to exploit the self-complexation properties of the cyclic carbohydrate in terms of solubility and reduced catabolism, although the relatively large carrier was expected to impair recognition processes at the molecular level. In fact, direct linkage of the enkephalin analog DPDPE to β-cyclodextrin was found to significantly reduce receptor affinity and selectivity [14]. Conversely, in recent studies we found that by the use of suitably sized spacers β-cyclodextrin/peptide aldehyde constructs retain full inhibitory potencies for cysteine proteinases [15]. Efficient synthetic routes for the production of mono- and hepta-functionalized β-cyclodextrin containing linear and flexible carboxyalkyl spacers of increasing chain length were therefore elaborated to investigate the role of the spacer in the presentation of protease inhibitors to the cyclodextrin template for recognition and binding by the enzymes [15, 16].

In view of the results obtained in these model studies, the tetra- and heptagastrin peptides [Nle^{15}]-HG-[14-17] and [Nle^{15}]-HG-[11-17] were grafted to β-cyclodextrin functionalized as mono-(6-succinylamino-6-deoxy)-derivative to produce the tetrapeptide (**1**) and heptapeptide (**2**) conjugate as shown in Fig. 1. Since β-cyclodextrin offers the additional option of a perfacial functionalization, the heptakis-[Nle^{15}]-HG-[14-17]/β-cyclodextrin conjugate (**3**) was synthesized (fig. 2) to analyze the effect of an *in loco* artificially enhanced concentration of the ligand on signal transduction efficacy.

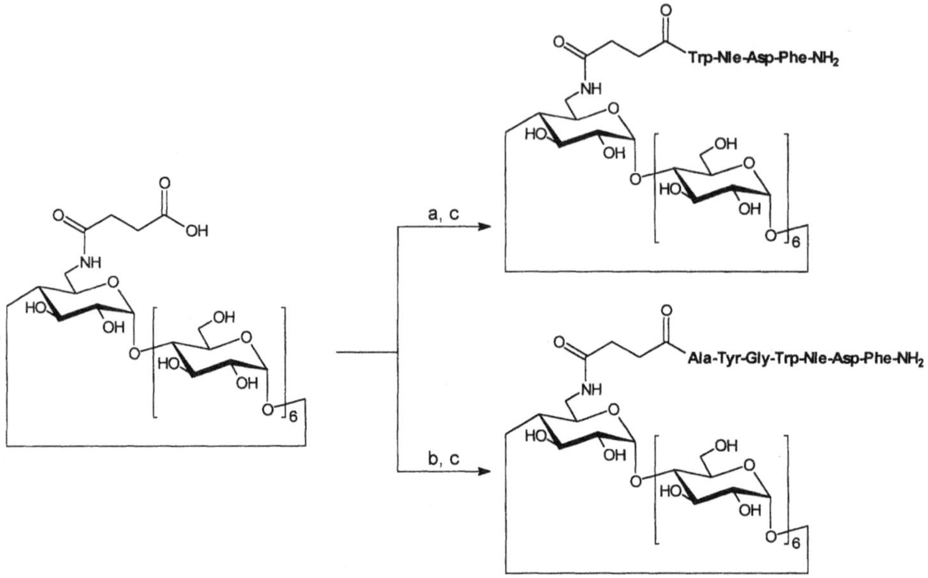

Fig. 1. Synthesis of gastrin peptides/β-cyclodextrin conjugates: a) H-Trp-Nle-Asp(OtBu)- Phe-NH$_2$/EDC/HOBt/DMF; b) H-Ala-Tyr(tBu)-Gly-Trp-Nle-Asp(OtBu)-Phe-NH$_2$/EDC/HOBt/ DMF; c) 90% TFA/1% 1,2-ethanedithiol.

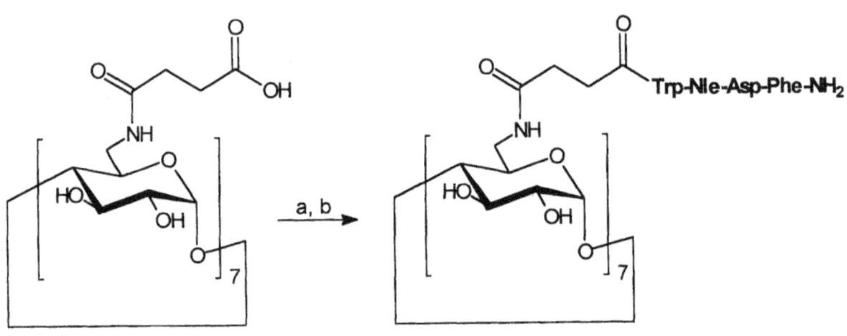

3

Fig. 2. Synthesis of the hepta-conjugate of the gastrin peptide [Nle^{15}]-HG-[14-17] with heptakis-(6-amino-6-deoxy)-β-cyclodextrin using the succinyl moiety as spacer: a) H-Trp-Nle-Asp(OtBu)-Phe-NH$_2$/PyBOP/NEt$_3$/DMF b) 95% aqueous TFA containing 1% 1,2-ethanedithiol.

^{1}H NMR conformational analysis of the tetrapeptide conjugate **1** in water clearly revealed a bended structure of the peptide backbone that is supported by the through-space NOEs from the aromatic side chains of both Trp and Phe as well as from the C-terminal amide protons of the carbohydrate moiety. Despite the lack of an ordered conformation such as a β- or γ-turn, the peptide moiety appears to be significantly restricted in its conformational space by self-complexation with the β-cyclodextrin carrier. This is further supported by the CD spectra (fig. 3) that show in the far UV a remarkably increased dichroism when compared to the unbound tetrapeptide, and in the near UV well resolved vibronic L_b bands of the phenyl and indole chromophores. Conversely, the NOEs of the heptapeptide conjugate **2** were of weak intensity suggesting a high degree of conformational freedom that is further confirmed by the CD spectra in the far UV region (fig. 4). In fact, the CD spectrum correlates well with a mainly unordered structure despite of the strong red shift of the negative maximum. In the near UV region the partly resolved negative band centered around 280 could derive from a restricted conformational space of the Tyr side chain, a possible result of interactions with the β-cyclodextrin moiety as suggested by the weak NOEs between the aromatic protons of Tyr and β-cyclodextrin. A shielding of the Trp and Phe side chains by the carrier in conjugate **1** correlates well with its increased stability toward enzymatic digestion with chymotrypsin and Asp-N protease (fig. 5 and 6).

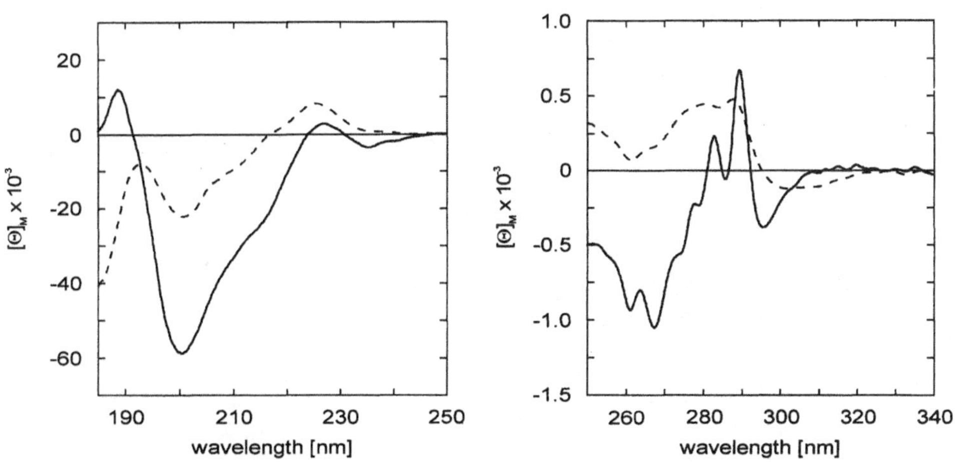

Fig. 3. CD spectra of [Nle[15]]-HG-[14-17]/β-cyclodextrin (1) (——) and Ac-Trp-Nle -Asp-Phe-NH₂; (-----) in 5 mM phosphate buffer (pH 7.0) at 20 °C in the near and far UV.

The host-guest complexation that was clearly observed for tetragastrin is not expected to interfere with binding of the ligand to the receptor since the stability of such complexes is known to be relatively low. As shown in Table 1, the succinyl spacer appears to be sufficient to guarantee recognition of the tetragastrin moiety. Sterical interference of the bulky (β-cyclodextrin moiety is not critical since even the signal transduction process is largely retained. With insertion of the tripeptide Ala-Tyr-Gly as additional spacer between message sequence and template the binding affinity was remarkably enhanced and full hormonal potency was recovered.

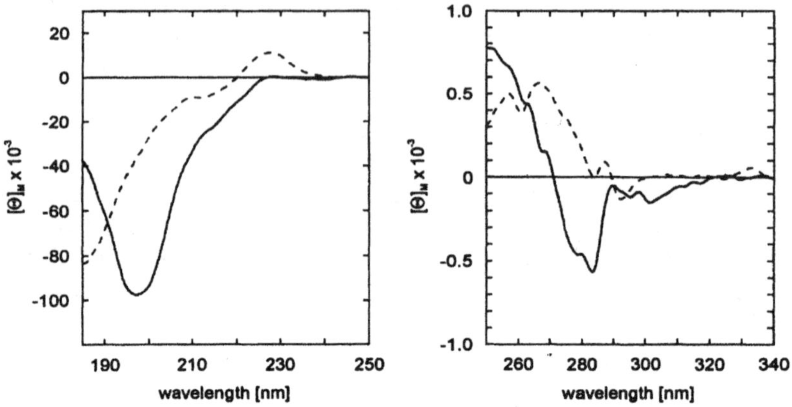

Fig. 4. CD spectra of [Nle^{15}]-HG-[11-17]/(β-cyclodextrin (2) (___) and Ac-Ala-Tyr- Gly-Trp-Nle-Asp-Phe-NH$_2$ (-----) in 5 mM phosphate buffer (pH 7.0) at 20 °C in the near and far UV.

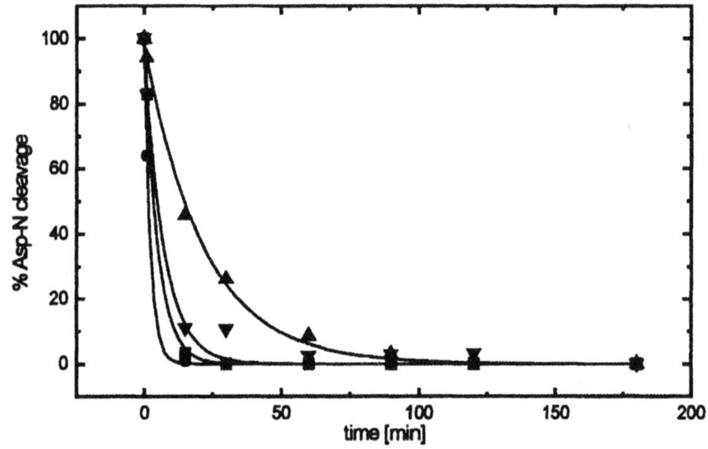

Fig. 5. Chymotryptic digestion of the conjugate 1 (?), 2 (¦), as well as of the unbound peptides Ac-Trp-Nle-Asp-Phe-NH$_2$ (?) and Ac-Ala-Tyr-Gly-Trp-Nle-Asp-Phe-NH$_2$ (?).

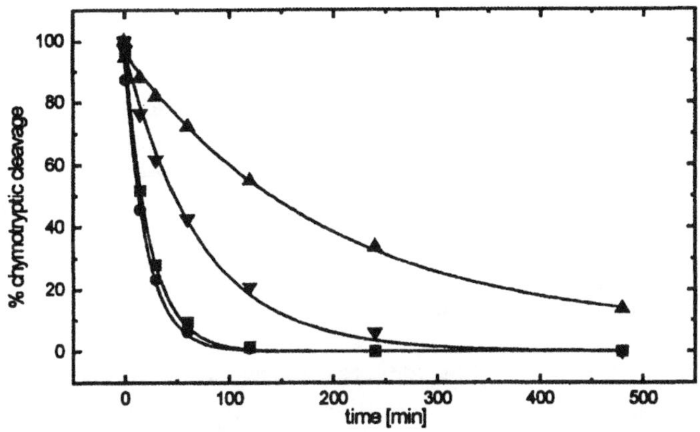

Fig. 6. Digestion of the comjugate ***1*** *(▲) and* ***2*** *(■) as well as of the unbound peptides Ac-Trp-Nle-Asp-Phe-NH$_2$ (▼) and Ac-Ala-Tyr-Gly-Trp-Nle-Asp-Phe-NH$_2$ (●) by the Asp-N-protease.*

Table 1. Binding affinities (IC$_{50}$) of gastrin peptides and (-cyclodextrin conjugates to CCK-B/ gastrin receptors overexpressed in CHO cells and inositol phosphate production (EC$_{50}$)

Ligands for CCK-B/gastrin receptor	IC$_{50}$ [nM]	IC$_{50}$ [nM]
[Thr, Nle]-CCK-9	0.33	0.09
Ac-Trp-Nle-Asp-Phe-NH$_2$	4.8	1.7
Pyr-Ala-Tyr-Gly-Trp-Nle-Asp-Phe-NH$_2$	1.6	0.12
(-CD-NH-CO(CH$_2$)$_2$CO-Trp-Nle-Asp-Phe-NH$_2$ (1)	26.3	1.53
(-CD-NH-CO(CH$_2$)$_2$CO-Ala-Tyr-Gly-Trp-Nle-Asp-Phe-NH$_2$ (2)	5.7	0.11
(-CD-[NH-CO(CH$_2$)$_2$CO-Trp-Nle-Asp-Phe-NH$_2$]$_7$ (3)	25.4	17.3

The 1D NMR spectrum of the heptaconjugate 3 in water showed strong broadening of the signals and thus was consistent with a system of multiple conformers deriving from fast interconversion of peptide/peptide and/or peptide/carrier interactions. These reciprocal interferences lead to significantly reduced receptor binding affinities per unit of tetrapeptide and in an even more pronounced manner, to reduction of receptor activation potency.

By comparing the binding affinities of the unbound and cyclodextrin-grafted gastrin peptides it is evident that for the conjugate 2 recognition by the receptor is identical to that of the tetrapeptide and that the Ala-Tyr-Gly sequence portion allows the message site of the gastrin to reach the functional binding site and fully activate the receptor despite the bulky cyclodextrin plug. In fact, docking experiments onto the homology modeling-derived structure of the human CCK-B receptor [10] clearly revealed that only with the heptagastrin conjugate 2 can the C-terminal tetrapeptide fully occupy the binding pocket identified with the lipo-gastrin derivatives [9]. From a 500 ps MD simulation in a triphasic solvent system a picture was obtained of the β-cyclodextrin moiety protruding from the protein surface into the extracellular aqueous compartment. It involved in a fluctuating interaction pattern of various hydroxyl groups of the carbohydrate with charged and hydrophilic side-chain functionalities of protein loop residues. In this docking model the peptide moiety is intensively engaged in intermolecular interactions involving residues of the extracellular loops as well as of the transmembrane helices, in good agreement with the previously proposed interaction mode [9].

Because of the ability of (β-cyclodextrin to modulate penetration of ligands into binding clefts it proved to be an interesting tool for validating our CCK-B receptor model. However, this template, which can be hetero-bis-functionalized, could also serve as a drug vector that contains in addition to the address labels for specific targeting of cells, (e.g. via selective proteinase inhibitor or peptide hormones) covalently grafted drugs for medicinal purposes.

Acknowledgments

The study was partly supported by SFB 469 (grant A-2/Moroder/ Machleidt) and SFB 266 (grant B-9/Kessler/Moroder).

References

1. Wank, S.W., Am. J. Physiol., 269 (1995) G628.
2. Baldwin, J.M., Curr. Opin. Cell Biol., 6 (1994) 180.
3. Strader, C., Fong, T., Tota, M., Underwood, D. and Dixon, R., Annu. Rev. Biochem., 63 (1994) 101.
4. Underwood, D.J. and Prendergast, K., Chem. Biol., 4 (1997) 239.
5. Romano, R., Musiol, H.J., Weyher, E., Dufresne, M. and Moroder, L., Biopolymers, 32 (1992) 1545.
6. Romano, R., Dufresne, M., Prost, M.C., Bali, J.P., Bayerl, T.M. and Moroder, L., Biochim. Biophys. Acta, 1145 (1993) 235.
7. Moroder, L., Romano, R., Guba, W., Mierke, D.F., Kessler, H., Delporte C., Winand, J. and Christophe, J., Biochemistry, 32 (1993) 13551.
8. Lutz, J., Romano-Götsch, R., Escrieut, Ch., Fourmy, D., Mathä, B., Müller, G., Kessler, H. and Moroder, L., Biopolymers, 41 (1997) 799.
9. Gurrath, M., Müller, G., Höltje, H.D. and Moroder, L., In Kungl, A.J. and Andrew, P.J. (Eds.) Proceedings of the 2nd International Conference on Molecular Structural Biology, Vienna, p. 124,
10. Tracy, H.J. and Gregory, R.A., Nature, 204 (1964) 935.
11. Moroder, L., J. Peptide Sci. 3 (1997) 1.
12. (Schaschke, N., Fiori, S., Weyher, E., Escrieut, C., Fourmy, D., Müller, G. and Moroder, L., J. Am. Chem. Soc., 1998, unpublished).

13. Djedaini-Pilard, F., Désalos, J. and Perly, B., Tetrahedron Lett., 34 (1993) 2457.
14. Hristova-Kazmierski, M.K., Horan, P., Davis, P., Yamamura, H.I., Kramer, T., Hoevath, R., Kazmierski, W.M., Porreca, F. and Hruby, V.J., Bioorg. Med. Chem. Lett., 3 (1993) 831.
15. Schaschke, N., Musiol, H.J., Assfalg-Machleidt, I., Machleidt, W., Rudolph-Böhner, S. and Moroder, L., FEBS Lett., 391 (1996) 297.
16. Schaschke, N., Musiol, H.J., Assfalg-Machleidt, I., Machleidt, W. and Moroder, L., Bioorg. Med. Chem. Lett., 7 (1997) 2507.

The C-terminus of δ-opioid receptor does not control DPDPE-induced desensitization

De-He Zhou [a], Chun-He Wang [a], Qiang Wei [a], Gang Pei [b] and Zhi-Qiang Chi [a]

[a]Shanghai Institute of Materia Medica; [b]Shanghai Institute of Cell Biology, Chinese Academy of Sciences, Shanghai, 200031, China

Introduction

δ, μ and κ opioid receptors are members in the G-protein coupled receptor (GPCR) family and negatively regulated the cellular cAMP system. Recently successful cloning of the receptor cDNAs greatly facilitated further studies on their structure and function relationship in the exploration of the elusive mechanism of opioid action [1].

Previously mutation studies on other GPCRs have proved that their C-terminals are important in the receptor regulation process after sustained agonist exposure [2]. For opioid receptors, the roles of their C-terminals in agonist-dependent receptor sequestration and down regulation have also been defined. However, the contributions of opioid receptors C-terminals in receptor activation and desensitization, which are important to the development of tolerance, have not been thoroughly elucidated.

In our recent work, the 31 amino residues of C-terminal truncated δ opioid receptors and wild-type were expressed stably in Chinese Hamster Ovary (CHO) cells [3] and the roles of the C-terminal of the δ-opioid receptor in the agonist-dependent G-protein activation and desensitization were discussed.

Results and Discussion

To rule out the effects of receptor density on receptor function, clones expressing different numbers of truncated and wild-type δ-opioid receptors were employed. Clone CHO-T stably expressed truncated receptors at a high level; CHO-t expressed truncated receptors at a relatively low level; CHO-W expressed high levels of the wild type and CHO-w expressed wild type at a relatively low level. Saturable [^{3}H]Diprenorphine binding and Scatchard analysis were done to determine receptor numbers and characterize receptor binding properties (table. 1).

There was no significant differences between the K_d values of truncated receptors and wild-types, both of which were similar to those reported for δ-opioid receptors.

Table 1.The K_d and B_{max} of the [^{3}H]Diprenorphine binding to clone CHO-T, CHO-t, CHO-W, and CHO-w.

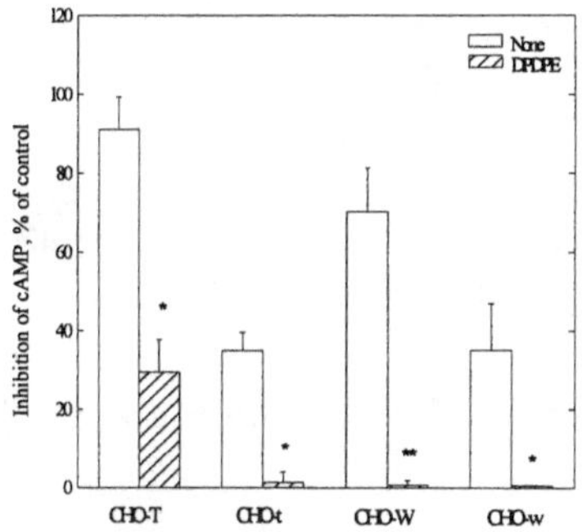

Fig. 1. DPDPE induced receptor desensitization in inhibiting the forskolin-stimulated cellular cAMP production.

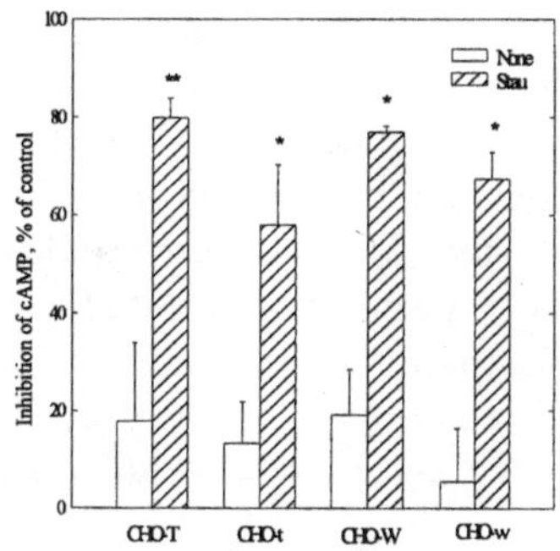

Fig. 2. The effect of staurosporine(Stau) on the DPDPE-induced desensitization of the CHO-T, CHO-t,CHO-W and CHO-w.

Table 2.The EC_{50} values of DPDPE in stimulating the specific [^{35}S]GTPγS binding to the membrane fragments of clone CHO-T, CHO-t, CHO-W, and CHO-w.

CLONES	Receptor expressed	K_d(nM)	B_{max}(pmol/g protein)
CHO-T	truncated δ-opioid receptor	1.35±0.82	4150±1180
CHO-t	truncated δ-opioid receptor	1.67±0.65	65.0±24.0
CHO-W	wild-type δ-opioid receptor	1.27±0.99	4330±300
CHO-w	wild-type δ-opioid receptor	1.95±0.98	410.4±144.0

To determine whether the C-terminal of the δ-opioid receptor is involved in coupling with the G-protein, [^{35}S] GTPγS binding assay was performed. When membrane fragments of CHO-T, CHO-t, CHO-W, and CHO-w were incubated with the typical δ-selective agonist [D-Pen^2, D-Pen^5] enkephalin (DPDPE), dose-dependent increases of the specific [^{35}S] GTPγS binding were observed on all clones. EC_{50} values of DPDPE on all clones were very similar (table 2), indicating that the C-terminal truncation of δ-opioid receptor did not affect the activation of the G-protein.

It has long been suspected that the phosphorylation of the third cytoplasmic loop or the C-terminal is responsible for desensitization. To determine whether the C-terminal of the δ-opioid receptor is needed in agonist-dependent receptor desensitization, DPDPE-induced desensitization of the C-truncated δ-opioid receptor was analyzed by cAMP assay. As plotted, DPDPE could greatly inhibit the forskolin-stimulated increase of cellular cAMP on cells without agonist pretreatment, but hardly affected cells, pretreated with 1 μM DPDPE for 10 min (fig. 1). The same results were obtained from wild types. Because the δ-opioid receptor with deletion of the C-terminal containing phosphorylation sites still underwent receptor desensitization, it can be concluded that phosphorylation of the C-terminal is not crucial for agonist-dependent desensitization. It is interesting to study and evaluate whether phosphorylation is involved in the desensitization of the truncated form. In our experiment, a potent nonspecific protein kinase inhibitor, Stauosporine (Stau), which could block the desensitization of the δ-opioid receptor in neuronal cells was employed. When CHO-T, CHO-t, CHO-W and CHO-w were incubated in medium containing 1 μM Stau at 37 °C for 15 min, desensitization induced by DPDPE could no longer be observed on either type of receptor (fig. 2). The vulnerability of desensitization to a protein kinase inhibitor suggests that agonist-stimulated desensitization of the C-truncated δ-opioid receptor is still a phosphorylation related process which might occur on the receptor but not the C-terminal. To define domains and their roles associated with the receptor desensitization process thoroughly, more detailed research should be performed.

Acknowledgments

This work was supported by the National Natural Science Foundation of China (No39630350) and the National Commission of Science & and Technology of the People's Republic of China.

References

1. Kieffer, B.L., Cell. Mol. Neurobiol., 15 (1995) 615.
2. Yu, S.S., Lefkowitz, R.J. and Hausdorff, W.P., J. Biol. Chem., 268 (1993) 337.
3. Wang, C.H., Zhou, D.H., Chen, J., Li, G.F., Pei, G. and Chi, Z.Q., Acta Pharmacol. Sin., 18 (1997) 337.

Session G
Peptides in therapeutics and diagnosis

Chairs: Yu-Cang Du
Shanghai Institute of Biochemistry
Chinese Academy of Sciences
Shanghai, China

Arnold Satterthwait
Department of Molecular Biology
The Scripps Research Institute
U.S.A

Bing-Gen Ru
Department of Biochemistry and Molecular biology
College of Life Sciences
Peking University
Beijing, China

Detection of gene expression product of transgenic tobacco by antigenic peptide

Jia-Xi Xu [a*], Yuan Ma [b], Mi Ma [c] and Zhong-Ping Lin [d]
[a]*Department of Chemistry, Peking University, Beijing, 100871;*
[b]*Department of Chemistry, Tsinghua University, Beijing, 100084;*
[c]*Institute of Botany, Academia Sinica, Beijing 10044;*
[d]*Department of Biology, Peking University, Beijing, 100871, China*

Introduction

The need as well as the ability to cultivate special plants, such as antiviral wheat, anti-insect plants, economical plants etc, has increased with advances in biological science and technology and has created significant economic benefits. It is easy to detect DNA and RNA in transgenic processing. However, it is difficult to detect the protein as gene expression product due to low concentration and difficult separation and purification [1]. Now we have used a synthetic antigenic peptide to solve this problem. After predicting and synthesizing an antigenic peptide of isopentyl transferase in transgenic tobacco, a key enzyme in gene expression, a peptide with the sequence IHARQQEQKF was conjugated to BSA. Its antibodies were then obtained from rabbit sera after immunizing the rabbit with this conjugated antigen. The antibody can be used for the qualitative and quantitative determination of gene expression product in crude protein-extract of leaf, stem and root of transgenic tobacco by ELISA and Western blot methods.

Results and Discussion

Prediction of epitope
The nucleotide and amino acid sequences of the isopentyl transferase in transgenic tobacco was obtained from published data [2]. Its epitope was predicted according to the hydrophilicity (Hopp and Woods method), flexibility and accessibility by the computer program PC-Gene [3, 4]. Its secondary structure was predicted by the Chou and Fasman method [5]. Peptide IHARQQEQKF (212-221 amino acid residues) with high hydrophilicity, flexibility and accessibility was predicted.

Synthesis of epitopic peptides
The predicted peptide was synthesized by Merrifield solid phase peptide synthesis method with the acid-labile tert-butyloxycarbonyl (Boc) group for temporary protection and acid-stable groups for side chain protection [3, 4, 6]. Side chain-protected peptide-resin was cleaved by anhydrous hydrogen fluoride under anisole, 1, 2-ethandithiol and thioanisole as scavengers and washed with cooled ethyl ether. Peptide was extracted with 30% acetic acid and purified by gel filtration on Sephadex G10. It was further

purified by preparative HPLC and checked for homogeneity by reverse-phase HPLC. Its structure was confirmed by amino acid analysis and FAB-MS.

Immunologiccal methods

This antigenic peptide was conjugated to BSA by the glutaraldehyde method and its antibodies were then obtained from rabbit sera after immunizing rabbit with this conjugated antigen [7]. The antibody was used to detect the gene expression product of transgenic tobacco at a dilution of 1:132 with OD492 nm 0.95 by the ELISA method. The results are shown in Table 1. This is a simple method for the qualitative and quantitative determination of gene expression product in crude protein-extract of leaf, stem and root of transgenic tobacco using ELISA and Western blot methods.

Table 1.ELISA of isopentenyl transferase in transgenic tobacco.

Strain	E1-3			E2-1			CK		
	Leaf	Stem	Root	Leaf	Stem	Root	Leaf	Stem	Root
OD_{492}	0.18	0.32	0.91	0.20	0.31	0.90	0.06	0.08	0.06

OD_{492} >0.1, positive; OD_{492} <0.1, negative. E1-3, E2-1: transgenic tobacco strains. CK: non-transgenic tobacco strain, control.

References

1. Stiefel, V., Perez-Grau, L., Albericio, F., Giralt, E., Ruiz-Avila, L., Ludevid, M.D. and Puigdomenech, P., Plant Mol. Biol., 11 (1988) 483.
2. Barry, G.F., Rogers, S.G. and Fraley, R.T., Proc. Natl. Acad. Sci. USA, 81 (1984) 4776.
3. (Xu, J.X. and Cai, M.S., Chin. Chem. Letts., 9 (1998), unpublished).
4. Xu, J.X., Cai, M.S. and Shi, Y.E., Chem. J. Chinese Univ., 17 (1996) 424.
5. Chou, P.Y., Fasman, G.D., Biochemistry, 13 (1974) 222.
6. Merrifield, R.B., J. Am. Chem. Soc., 85 (1963) 2149.
7. Craig, S., Millerd, A. and Goodchild, D.J., Aust. J. Plant Physiol., 7 (1980) 339.

Advantages of Calcitonin complexed with SA liposome

Die Wang [b], Chao Yu [a], Hai-Xing Xuan [b], Da-Fu Cui [a] and Qi-Shui Lin [b]
[a]Shanghai Institute of Biochemistry, Academia Sinica; [b]State Key Laboratory of Molecular Biology, 320 Yue Yang Road, Shanghai, 200031, China

Introduction

Polypeptide drugs interact with target cells, but there are concomitant side effects and toxicity to normal cells. Consequently, their applications and dosage usually have to be limited. Clinical application of peptide drugs must meet the requirements at low toxicity, potency enhancement and dosage reduction. Liposome is a successful example of Various carrier systems that are used to lower toxicity, enhance potency and prolong half-lives [1]. Drugs encapsulated in liposomes have been applied for treatment and substantial effects have been observed, e.g. lower toxicity or side-effects.

In the present study, SA liposome was used to improve the effect of peptide drugs *in vivo*. Favorable results were obtained, such as resist once to degradation, higher half-life, and better absorption. Cationic liposome is one kind of liposome with a positively charged surface. Peptide drugs can interact with cationic liposomes by been encapsulated inside the liposome. The positive charges on the surface of cationic liposomes can affect static electricity absorption with the negative charged segment of polypeptide drugs. Also, the hydrophobic polypeptide residue can insert into the lipid bilayer. Compared to normal liposomes, cationic liposome as a drug carrier, has the advantage of stronger affinity with the negatively charged cell surface, thus facilitating drug absorption. The SA liposome, which we invented, contains stearylmine and DOPE, both of which are naturally occurring lipids, and has low immunogenicity and high stability [2].

Calcitonin, an important peptide hormone secreted by thyroid gland C cells, regulates calcium metabolism and skeletal metabolism *in vivo* [3]. Due to its conspicuous biological activity which can be measured sensitively, and its stability, we selected sCT (salmon calcitonin) and two kinds of hCT (human calcitonin) analogues as peptide model to study the interaction with SA liposome.

Methods

1. Preparation of SA liposomes
Preparation of SA liposomes was performed according to the previous report [4].

2. Solid phase synthesis of hCT-I, II and their purification
The peptide was synthesized step by step by solid phase method in the direction of sequences from C-terminal to N-terminal. The N-terminal and the peptide side chains were protected by Boc group. The peptide was then cleaved from the resin by HF,

desalted and purified by RP-HPLC. Mass spectrum analysis and N-terminal sequencing of the peptide was in accordance with the theoretical values.

3. Preparation of SA liposomes-Calcitonin complex

1) Freezing-thawing pack:

Calcitonin and SA liposome were diluted separately with physiologic saline and then mixed. The mixture was left at room temperature for 5 minutes, then subjected to freezing-thawing cycle with liquid nitrogen—37 °C 3 times.

2) Electrostatic adsorption method:

Calcitonin and SA liposome were diluted separately with physiologic saline and then mixed. The mixture was left at room temperature for 25 -30 minutes.

4. Biological assay for hypocalcemic activity of free calcitonin and SA-calcitonin complex

The hypocalcemic activity of free calcitonin and SA-calcitonin complex were measured as follows: Male rats (Wistar) weighing 80 grams were fasted overnight (about 18 hours), but fed with ddH_2O. Random grouping was according to rat weight. Each group was composed of eight rats. Six control rats were injected with physiologic saline and SA liposome. Free salmon calcitonin (sCT), free human calcitonin analogue I or II (hCT-I, hCT-II) and SA-sCT/SA-hCT-I, II complex were injected intra-peritoneally in respective groups. Each injection was 0.4 ml per rat, which included 2.5 mg SA liposome or 20 ng sCT or 50 ng hCT-I, II. The serum calcium concentrations at different times after injection were analyzed by the OCPC method [5].

Results

1. *In vivo* half-life of free sCT and SA-sCT complex

After 1 hour of free sCT injection, the concentration of rat serum calcium reached the lowest value and the calcium-lowering activity could be maintained for 4 hours. However, when SA-sCT complex was injected, the hypocalcemic activity was not as strong as free sCT at 1 hour, but at 2 hours was almost the same, and some calcium-lowering potency was maintained even after 4 hours.

2. *In vivo* half-life of free human calcitonin analogue I (hCT-I) and SA-hCT-I complex

After 1 hour of free hCT-I injection, the concentration of rat serum calcium reached the lowest value and returned to pre-injection levels at 5 hour. For SA-hCT-I complex, although the hypocalcemic activity at 1 hour was a little lower than that of free hCT-I, it was almost the same at 2 hour. Results showed that when SA-hCT-I complex was injected the serum calcium level after 2 hours was obviously lower than that of free hCT-I injection ($P<0.005$) and some Ca-lowering activity was maintained even after 5 hours.

3. Biological activity of SA-hCT-II complex formed by freezing-thawing or electrostatic adsorption method

After 4 hours of injection the free hCT-II was completely degraded. The concentration of rat serum calcium reached the lowest level at 2 hour after injection of SA-hCT-II complex formed by freezing thawing and returned to normal at 6 hour. For those SA-hCT-II complexes formed by electrostatic adsorption method, the hypocalcemic activity was similar to that formed by freezing thawing in the first 1~2 hours, but was higher after 2 hours. The difference was significant, P value < 0.01.

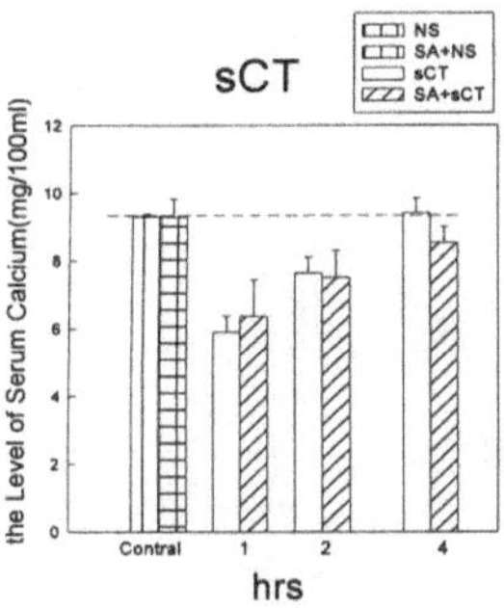

Fig. 1. The time-dose curve of sCT and SA-sCT complex or electrostatic adsorption.

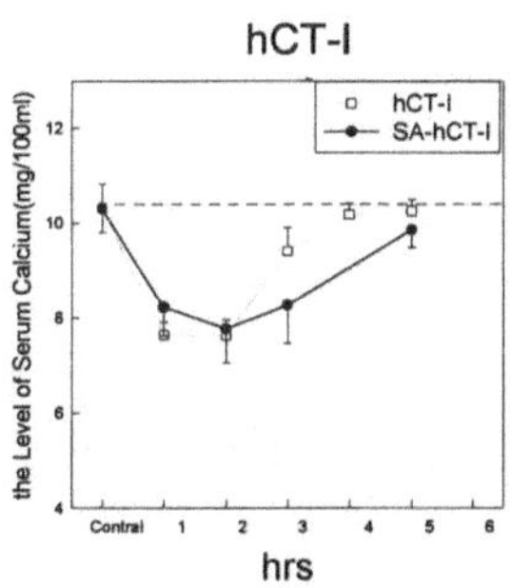

Fig. 2. The time-dose curve of hCT-I and SA-hCT-I complex.

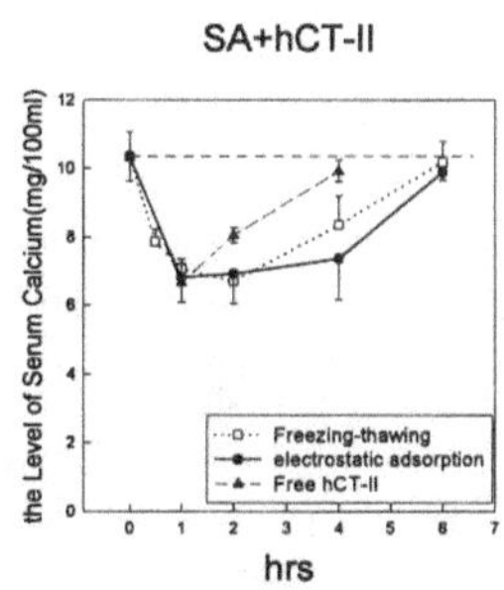

Fig. 3. The dose-effect curve of SA-hCT-II complex formed by freezing-thawing pack.

Discussion

As a peptide drug carrier, liposomes possess the following advantages: 1). liposomes are composed of natural occurring lipids; they can be biodegraded; they do not have toxicity and immunogenicity. 2). Both water-soluble and fat-soluble drugs could be embedded or inserted in liposomes, and then gradually released, so that activity can be sustained for a longer period. 3). The interaction of liposomes with cells through endocytosis or fusion can deliver drugs directly into cells, thus avoiding the necessity of using high drug concentrations. 4). The distribution of drugs in vivo could be controlled to certain extent and drug release could be directed to pathologically changed tissues if the liposome was specially prepared for targeting, thus reducing side effects.

Liposomes have been widely applied for the delivery of small molecular drugs. The most important benefits are that they (a). stabilize the peptide from degradation; (b).prolong the half-life of drugs *in vivo* and (c). Possess slow-released action. Compared with other carrier systems, liposomes do not need covalent cross linkage with drugs, and the chemical structure of the drugs is not modified. The operation of cationic liposome is even simpler than conventional liposomes. The complex formed by positive charges on the surface of cationic liposomes and polypeptide drugs can develop an affinity with the negatively charged cell surface, which favors drug absorbance or release.

The short half-life of polypeptide drugs requires repeated administration. At the same time a high concentration would result in certain side effects. The hypocalcemic activity of free calcitonin reaches its maximum in 1 hour after injection, and the effect will dissipate after 4 hours. However, the calcitonin-SA cationic liposome complex leads to a slow action of calcitonin in vivo and the peak value of hypocalcemic activity is reached between 1~2 hours, indicating the prolonged half-life. The drugs retained hypocalcemic activity even after 5~6 hours of injection.

The results indicate that the complex formed by SA cationic liposome with peptide drugs might be good and worth to be further exploration.

References

1. Langer, R., Nature, 392 (1998) 5.
2. Wang, D., Jing, N.H. and Lin, Q.S., Biochem. Biophys. Res. Commun., 226 (1996) 450.
3. Hirsch, P.F., Science, 146 (1964) 412.
4. Lin, Q.S., Yang, J.P. and Wang, D. (Eds.) Biopolymers and Bioproducts: Structure, Function and Applications (Published for the Organizing Committee, 11th FAOBMB Symposium) Samakkkhisam (dokya) Public Company Limited, 1995, p. 46-53.
5. Xue, J.Z. and Gao, X.J., Chinese Journal of Medical Laboratory Technology, 7 (1984) 147.

Hypoglycemic effects of insulin sublingual drops containing azone on rats and rabbits

Hong Peng, Qiu-Hua Gao, Qun Chen, Yu-Shan Zhu and Kai-Xun Huang

Institute of Pharmacy, Huazhong University of Science and Technology, Wuhan, 430074, China

Introduction

Transmucosal delivery of insulin(e.g. rectal, nasal, buccal, sublingual) has been shown to be useful, but enhancers are necessary to improve bioavailablities. Azone is a novel enhancer for transdermal delivery of pharmaceuticals, and some studies show that it is also an effective enhancer for nasal and rectal absorption of insulin [1, 2]. 1, 3-Propanadiol, a polar solvent and transdermal promoter, can improve the concentration of Azone in stratum corneum, and be synergistic with Azone to increase transdermal absorption of pharmaceuticals. Because the buccal membrane is a stratified squamous epithelium, similar to skin, we investigated hypoglycemic effects of insulin sublingual drops in combination with Azone, 1, 3-propandiol and Tween80 on rabbits and rats in this study. The sublingual absorption of insulin then was estimated from the total decline of glucose levels compared to control.

Materials and Methods

Materials

Pork insulin powder (1 mg/27.2 u) was obtained from Xuzhou Biochemical Pharmaceutical Company. All chemicals used were of analytical grade and purchased from local sources. Insulin sublingual drops were prepared by dispersing the required amounts of insulin in pH7.4 PBS in combination with Azone, 1, 3-propanadiol and Tween 80. For subcutaneous administration, insulin was dissolved in pH 7.4 PBS.
Japan white rabbits and SD rats were purchased from the Center for Experimental Animals at Wuhan Tongji Medical University.

Methods

Female and male rabbits weighing 2.5 kg were fasted for 12 h prior to the experiment with free access to water. The rabbits were divided into 3 groups of 5 rabbit each: (1) control (sublingual administration of saline); (2) insulin sublingual drops group (sublingual administration of 2 u/kg insulin); (3) SC group (subcutaneous administration of 0.4 u/kg insulin), respectively.

Female and male rats weighing 200 g were fasted for 10 h prior to the experiment with free access to water. The rats were divided into 3 groups of (6 rats each), they are (1) control (sublingual administration of salt water); (2) insulin sublingual drops group (sublingual administration of 4 μ/kg insulin); (3) SC group (subcutaneous administration of 0.8 μ/kg insulin), respectively.

The rabbits were anesthetized with an i. v. dose of 40 mg/kg sodium pentobarbital, and the rats were anesthetized with an i. p. dose of 55 mg/kg sodium pentobarbital. The esophagus was surgically ligated to prevent swallowing of the dosing solution. A predose blood sample was taken and insulin was administered using a microliter syringe. Ten minutes later, the esophagus was released. Serial blood samples were collected at 0, 0.5, 1, 1.5, 2, 2.5, 3.0, 3.5, 4.0 h by cutting the ear vein of rabbits or cutting the tip of the tail of the rats. Glucose levels were immediately determined with test strips and Accutrend alpha blood glucose monitor.

Results and Discussions

Table 1.Glucose change (mmol/l) vs time after administration of insulin sublingual drops to rabbits ($\overline{X} \pm SD$).

Time (min)	Control (Sub. Saline)	Sublingual Drops (Sub. 4u/kg insulin)	SC Group (SC 0.8u/kg insulin)
0	6.9±1.0	7.0±1.3	6.8±0.7
30	7.0±0.8	4.1±0.8	4.0±0.8
60	6.5±0.9	3.1±1.4	1.8±1.0
90	5.9±0.7	2.8±0.9	1.7±0.8
120	6.4±1.2	2.9±1.0	1.4±0.8
150	6.9±1.6	3.0±1.0	1.7±1.0
180	6.7±1.3	3.1±0.8	2.0±1.2
210	7.1±1.5	3.4±1.4	2.0±1.2
240	6.9±1.0	3.5±1.0	3.1±2.0
AUC (mmol.min/l)	1674.2	817.5	585
D%		51.2%	65.1%

$D\% = (AUC_{control} - AUC_{Sub.})/AUC_{control} \times 100\%$

Hypoglycemic effects of insulin sublingual drops on rabbits

By comparison to control, insulin sublingual drops resulted in significant decreases in glucose levels ($P<0.05$, table1), The maximal glucose decrease for rabbits by sublingual absorption of insulin was 60.0±12.8% at 1.5 h; the average total decrease of glucose levels compared to control (D%) was 51.2%, and the relative bioavalability for rabbits was 15.7% (the bioavalibility=$(AUC_{control}-AUC_{Sub})/(AUC_{control}-AUC_{SC}) \times (Dose_{SC}/Dose_{Sub}) \times 100\%$, and AUC represents the average area under the change in blood glucose vs time curve).

Hypoglycemic effects of sublingual drops on rats

Insulin sublingual drops could also produce significant hypoglycemic effects on rats comparing to control ($P<0.05$, table 2), The maximal glucose decrease by sublingual absorption of insulin was 53.2±14.3% at 1.0 h, the average total decrease of glucose levels compared to control (D%) was 42.9%, and the relative bioavailability for rats is 19.6%.

Table 2.Glucose change (mmol/l) vs time after administration of insulin sublingual drops to rats ($\overline{X} \pm SD$).

Time (min)	Control (Sub. Saline)	Sublingual Drops (Sub. 4u/kg insulin)	SC Group (SC 0.8 u/kg insulin)
0	4.8±0.5	4.7±0.4	4.8±0.3
30	5.0±0.9	3.0±0.6	3.1±0.7
60	4.9±0.9	2.2±0.7	2.1±0.6
90	5.1±1.0	2.3±0.5	2.0±0.7
120	5.2±1.3	2.4±0.5	2.3±0.8
150	4.8±0.9	2.7±0.4	2.6±0.9
180	4.9±1.5	2.8±0.9	2.9±1.0
210	4.7±0.6	3.0±0.8	3.2±0.8
240	5.1±0.6	3.6±1.0	3.5±1.0
AUC (mmol.min/l)	1182	675	663.8
D%		42.9%	43.8%

$^*D\% = (AUC_{control} - AUC_{Sub.})/AUC_{control} \times 100\%$

These results for rabbits and rats show that the combination of Azone, 1,3-propanadiol and Tween 80 promotes insulin absorption from buccal mucosa effectively, and may be a novel enhancer for sublingual absorption of insulin, probably by a synergistic effect between Azone and 1, 3-propanadiol.

References

1. Tessumi, I., Kazuya, A., Drug Delivery Syst., 7 (1992) 91.
2. Ritschel, W.A., Ritschel, G.B. and Ritschel, B.E.C., Clin. Pharmacol., 10 (1988) 645.
3. Cheng, G.S., Gong, S.J. and Zhou, R.R., Chinese Pharmaceutical Journal, 29 (1994) 467.

Session H
Other research related to peptides and proteins

Chairs: James P. Tam

Department of Microbiology and Immunology
Vanderbilt University
Nashville, USA

Subro Aimoto

Institute for Protein Research
Osaka University
Osaka, Japan

and

Xiao-Jie Xu

Department of Chemistry
Peking University
Beijing, China

Secretory expression of porcine insulin precursor in methylotrophic yeast *Pichia pastoris*

Yan Wang, Zheng-He Liang, You-Min Feng and You-Shang Zhang
Shanghai Institute of Biochemistry, Academia Sinica, 320 Yueyang Road, Shanghai, 200031, China

Introduction

The methylotrophic yeast *Pichia pastoris* is a recently developed host for heterologous protein expression. As nonconventional yeast, it has the following advantages over *S. cerevesiae*: readiness of high-density growth and scaling-up, and the presence of a powerful methanol-regulated promoter AOX1 (expressing AOX to 30% of total cell protein) [1]. Furthermore, the transformed strains with integrative expression are stable over generations. This system has been used in the expression of many heterologous proteins.

We have already expressed insulin precursor in *S. cerevisiae* [2] and *K. lactis* [3]. In order to increase the expression level, we have tried to express it in this system.

Results and Discussion

Gene Cloning

Porcine insulin precursor (PIP) gene with a spacer at the 5'-end obtained by PCR was inserted into the integrative plasmid pPIC9 just behind the KEX2 cleavage site. This spacer was used in *S. cerevisiae* to ensure complete cutting of PIP by KEX2 [4]. DNA sequencing of the constructed plasmid showed that the gene had the right sequence and was inserted into the right place.

Screening of Transformants with Multiple Integration

The plasmid was linearized by *Bgl II*, *Sal I* and *Sac I* separately to transform *P. pastoris* yeast GS115 using the spheroplast method. Transformants with three types of integration were obtained. With *Bgl II* double cutting, the linearized plasmid could transplace *AOX1* gene in the chromosome and form Mut^s (methanol-utilization slow) phenotype. It could also recircularize and integrate with a single crossover to form Mut^+ phenotype [5]. The phenotype of the strain was first examined by testing the growth on MD and MM. 17 Mut^s transformants appeared in 92 at a frequency of 18.5%, comparable to the overall frequency of Mut^s transformants (22%) reported previously [6]. With *Sal I* or *Sac I* linearization, all clones were Mut^+, confirming single crossover integration. Another testing method was shake tube culturing to determine growth differences after induction. The phenotype of the strain was further confirmed during induced expression.

In transformation experiments using the spheroplast method, the frequency of multiple integration is in general 1-10%. The expression normally correlates with gene dosage, so multiple integrated strains were screened.

The first method of screening multiple integrated transformants is measuring the expression of PIP with RIA (radioimmuno assay) of insulin. After 4 day induced expression, the culture supernatant was diluted and assayed by RIA. As shown in Table 1, the negative control p9 (cells transformed by pPIC9 without PIP gene) expressed no insulin precursor and several transformants were found to be false positive.

Table 1.RIA of some transformants.

Transformants	S11	B4	B5	B6	B24	**S11**	**B4**	**F50**	p9
RIAunit (μU)	24	27	21	21	0	18	19	18	0

(Culture supernatant 1:500 dilution, except **S11, B4** and **F50** 1:5000 dilution)
S: *Sal I* transformants; B: *Bgl II*; F: *Sac I*; p9: the negative control.

RIA is a good characterization method. But its drawback to quantification is that it has dilution effect, compared to negative control p9, the secretion of protein with insulin antigen activity was clean.

Another method for screening is dot blotting. After growing to saturation, equal numbers of cells were digested in tubes with lyticase (2 U/50 μl cell) at 37 °C for 4 hr and the digest collected was adsorbed on nylon membrane (Hybond-N). After denaturation, the membrane was hybridized with random primed, ^{32}P-labeled *PIP*. The data of dot blotting were consistent with those of RIA. As shown in Fig. 1, several clones with high copies were present.

Sal I transformants were screened by RIA. In 96 *Sal I* transformants, 20 were false-positive and 8 had multiple copies. *Sac I* and *Bgl II* transformants were screened mainly with dot blotting and checked with RIA. In 48 *Bgl II* transformants, 34 were false-positive and 4 had multiple copies. In 50 *Sac I* transformants, there were 10 false-positive and 10 multicopy transformants.

From the data, Sac I transformants had the highest frequency of multiple inserts, while Bgl II transformation gave transformants with higher copies. These data were obtained in one transformation, and further work must be done.

RIA showed that strains with multiple copies selected by dot blotting had a higher level of expression than those with a single copy. As shown in Table 1, the expression level of multicopy strains S11, B4 and F50 was almost 10 times higher than that of single copy strains. The expression was about 10 mg/L in shake flask culturing.

The third method was screening with G418. GS115 cells were transformed by pPIC9K with PIP gene and plated on MD. Cells on the plate were replated onto YPD and YPD containing 0.8mg/ml G418 in equal amounts. Clones on YPD containing 0.8 mg/ml G418 were at least 10 times less than those on YPD. Since G418 thus provided selective source for multicopy screening. Checking with dot blotting and RIA is still in progress.

SDS-PAGE

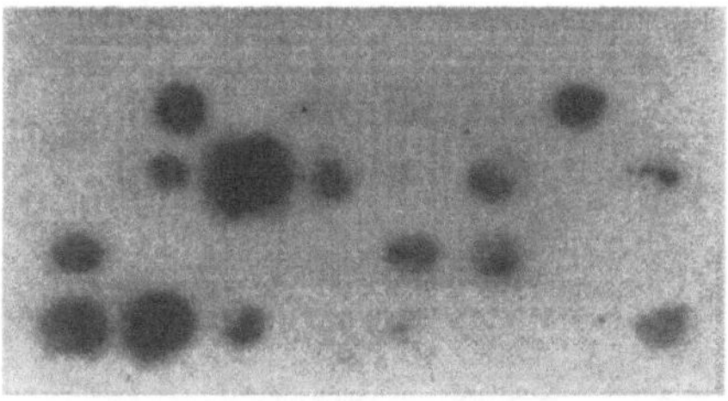

Fig. 1. One graph of dot blotting.(from left to right, down to up are S11, B4, B5, B24-B28, B29-B43 sequentially. The dot on the first up line right is positive control of diluted probe left to it is the negative control p9.)

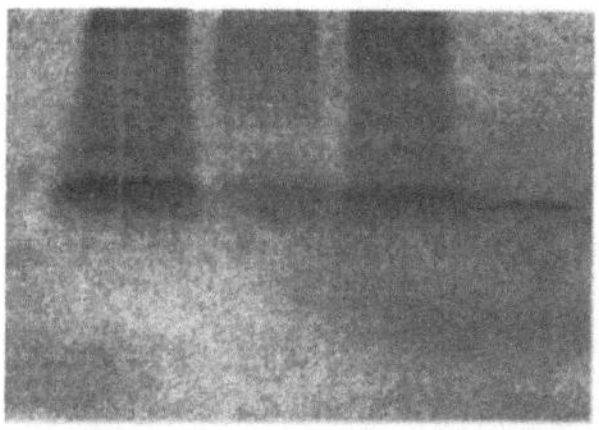

4 3 2 1
Fig. 2. SDS-PAGE (Tris-Tricine system). 1. PIP without spacer 2-4. Positive strains.

Tris-Tricine SDS-PAGE for small molecules was used to analyze the culture supernatant after induction. The sample was concentrated before loading. Positive strains showed a single band of correct MW, a little behind PIP without the spacer, while negative control p9 showed no band at this place (not shown).

Using several screening methods, we have got some strains with high copies. The high density fermentation of these strains will be done.

References

1. Cregg, J.M., Bio. Tech., 11 (1993) 905.
2. Zhang, Y.S., Science in China (Series C), 27 (1997) 1.
3. Feng, Y.M., Acta Biochimica et Biophysica Sinica, 29 (1997) 129.
4. Kjeldsen, T., Gene, 170 (1996) 107.
5. Clare, J.J., Bio.Tech., 9 (1991) 455.
6. Clare, J.J., Gene, 105 (1991) 205.

Chemo-enzymatic syntheses of eel calcitonin analogs having natural N-linked oligosaccharides

Toshiyuki Inazu[a], Mamoru Mizuno[a], Ikuyo Muramoto[a], Katsuji Haneda[a], Toru Kawakami[b], Saburo Aimoto[b] and Kenji Yamamoto[c]

[a]The Noguchi Institute, Tokyo; [b]Institute for Protein Research, Osaka University, Osaka; [c]Graduate School of Agriculture, Kyoto University, Kyoto, Japan

Introduction

Growing interest in cell surface glycoproteins as ligands having biologically important roles, such as cell recognition, and cell adhesion etc, has stimulated the development of new methods for convenient synthesis of glycopeptides. Recently, we reported the solid-phase synthesis of a glycopeptide containing the Asn (GlcNAc) residue by the dimethylphosphinothioic mixed anhydride (Mpt-MA) method without protecting the hydroxyl functions of the sugar moiety [1]. We also have reported the transglycosylation reaction of a synthetic peptide with the GlcNAc residue using endo-b-N-acetylglucosaminidase of Mucor hiemalis (Endo-M) [2]. We then tried to add the sugar chain to the protein without a natural sugar chain by the chemo-enzymatic method as a new strategy for glycoprotein synthesis. Eel Calcitonin (ECT) is a calcium-regulating hormone consisting of 32 amino acid residues, and has a consensus sequence "Asn-Leu-Ser" for N-glycosylation and no sugar chains. In this paper we describe the syntheses of eel calcitonin analogs 8, 9 and 10 having a natural N-linked sugar chain.

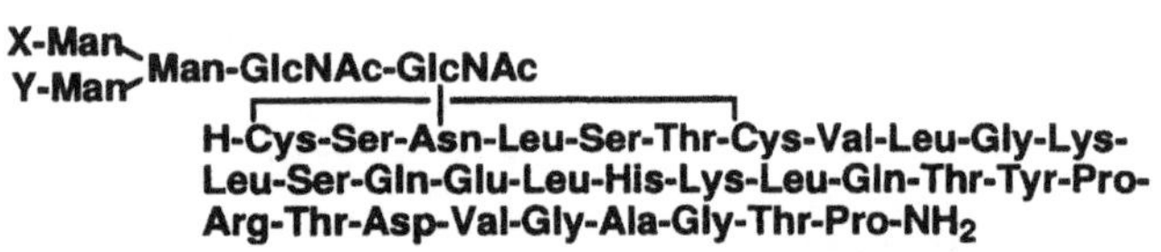

8: X, Y = NANA-Gal-GlcNAc **9**: X, Y = Gal-GlcNAc **10**: X=$(Man)_2$, Y=Man

Results and Discussion

In this work, our strategy for glycoprotein synthesis consisted of four steps. The first step is glycosylasparagine synthesis. The second is the synthesis of a glycopeptide moiety having N-acetylglucosamine using the Mpt-MA method. The third is a polypeptide moiety synthesis using the thioester segment condensation method. The last is a transglycosylation reaction using Endo-M. First, we prepared the glycosylasparagine derivative, which is the core unit of the N-glycopeptide. The N^{α}-Boc aspartic acid a-benzyl ester and N-acetylglucosaminyl azide derivative are dissolved in dichloromethane and tributylphosphine is added at - 78 °C [3]. The protected glycosylasparagine derivative was then obtained in 65% yield. Catalytic hydrogenation using palladium-carbon gave Boc-glycosylasparagine in quantitative yield.

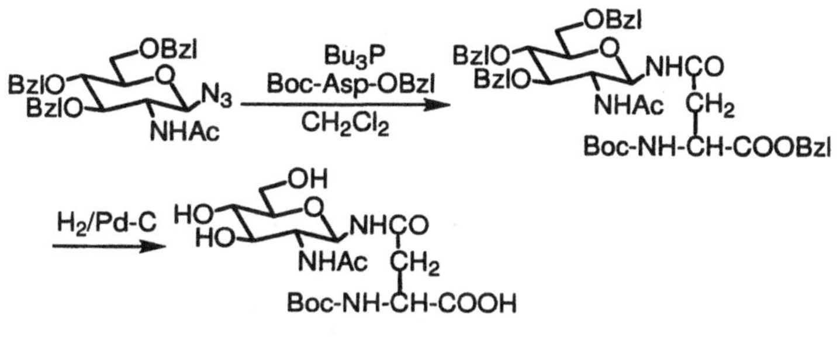

The glycosyl ECT having the Asn(GlcNAc) residue, [Asn(GlcNAc)3]-ECT (1), was synthesized by combining the Mpt-MA method and the thioester segment condensation method [4] on a solid support in good yield. The peptide thioester resin 3 was prepared from 2 by a Boc-strategy using the DCC-HOBt coupling method. The corresponding Mpt-MAs of Boc-Asn(GlcNAc)-OH, Boc-Ser(Bzl)-OH and Fmoc-Cys(Acm)-OH were introduced one by one into 3, and these couplings were repeated [1]. The N-terminal glycopeptide thioester segment 4 was obtained in 12% yield by treatment with anhydrous HF containing 10% anisole and reversed-phase HPLC (RP-HPLC). During the synthesis of 4 from the starting resin, no significant side reactions were observed. The C-terminal peptide segment 5 was prepared by a Boc-strategy. For the thioester method, the side-chain amino groups of lysine residues in segment 5 were protected by Boc groups. The glycopeptide thioester segment 4 and partially protected peptide segment 5 were added to a mixture of $AgNO_3$, 3-hydroxy-3, 4-dihydro-4-oxo-1,2,3-benzotriazine (HOObt) and DIEA in DMSO. The reaction mixture was stirred for 16 h at room temperature. The partially protected, ECT derivative was obtained in 78%. After cleavage of the Boc and Fmoc groups, and RP-HPLC purification, precursor 6 was prepared. The precursor 6 was treated with $AgNO_3$ and DIEA in aqueous DMSO, followed by 1N HCl/DMSO at room temperature to remove the Acm groups and form a disulfide bond [5]. After RP-HPLC purification, the glycopeptide analog of eel calcitonin 1 was obtained in 9% overall yield based on the number of amino groups in the starting NH_2-resin.

Next, the disialo and asialo complex-type oligosaccharide of egg yolk and the high mannose type oligosaccharide of ovalbumin were transferred to the GlcNAc moiety of 1 by Endo-M. The desired ECT analogs having a natural N-linked sugar chain 8, 9 and 10 were obtained in 8.5%, 7.5% and 3.5% yields, respectively. These conditions were similar to that described previously [2]. Endo-M very effectively transglycosylated the disialo complex-type oligosaccharide, which was the poorest substrate for hydrolysis.

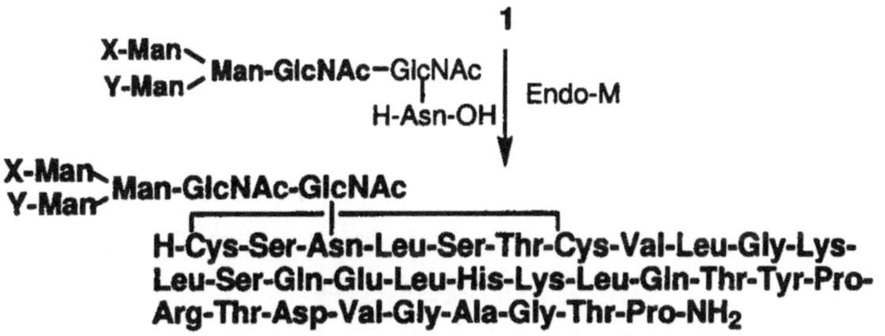

8: X, Y = NANA-Gal-GlcNAc **9**: X, Y = Gal-GlcNAc **10**: X=$(Man)_2$, Y=Man

References

1. Inazu, T., Mizuno, M., Kohda, Y., Kobayashi, K. and Yaginuma, H., In Nishi, N.(Ed.) Peptide Chemistry 1995, Protein Research Foundation, Osaka, 1996, p. 61.
2. Haneda, K., Inazu, T., Yamakoto, K., Kumagai, H., Nakahara, Y. and Kobata, A., Carbohydr. Res., 292 (1996) 61.
3. Mizuno, M., Muramoto, I., Kawakami, T., Seike, M., Aimoto, S., Haneda, K. and Inazu, T., Tetrahedron Lett., 39 (1998) 55.
4. Inazu, T. and Kobayashi, K. Synlett, 1993, 869; (Mizuno, M., Muramoto, I., Kobayashi, K., Yaginuma, H. and Inazu, T., Synthesis, unpublished).
5. Kawakami, T., Kogure, S. and Aimoto, S., Bull. Chem. Soc. Jpn., 61 (1996) 3331.

Design of specific inhibitors at the ATP binding site of DNA topoisomerases II

Liliane Assairi

Institut de Génétique et Microbiologie, URA 2225, Université de Paris-Sud, Centre d'Orsay, 91405 Orsay Cédex, France

Introduction

The activity of enzymes is regulated in various ways, one of which is by the structural changes occurring upon the phosphorylation and the dephosphorylation of proteins catalyzed by specific kinases, the synthesis and the activity of DNA topoisomerase II [1], an essential enzyme involved in the segregation of chromosomes occurring at the end of DNA synthesis. The enzyme acts as a dimeric protein and is also phosphorylated during the cell cycle. In order to follow this phosphorylation and afterwards variations in the regulation of enzymatic activity during the cell cycle, biosensors specific to either active form of the enzyme will to be designed and synthesized. For that purpose, the ATP recognition site of DNA topoisomerase II is analyzed. Two approaches are followed: the first consists of the determination of the amino acids involved in the recognition of ATP [2] and the second consists of the design and synthesis of new inhibitors [3, 4] which could map the topography of the ATP recognition site and constitute the basis for designing specific biosensors. These biosensors could specifically recognize one particular enzyme, and also the various conformational steps of the enzyme that occurs during a particular biological process.

Results and Discussion

The comparative analysis of the inhibitory effect at the ATP binding site of DNA topoisomerase II is shown in Table 1. The various of the inhibitory effect corresponds to an evolution of the topography of the ATP binding site of DNA topoisomerase II.

This variation of ATP hydrolysis is the consequence of the structural evolution of the genes coding for DNA topoisomerase II (fig. 1). Indeed, two separate genes code for the two subunits composing the *E. coli* although these two genes are adjacent in *B. subtilis*. Finally, only one gene codes for the whole monomer in eukaryotes. Thus, a fusion between the two genes which occurred through evolution, lead to modification of ATPase activity through the protein folding step.

Table 1.Inhibition of the ATP hydrolysis catalyzed by DNA topoisomerases II.

inhibitors	bacteria [1]	archaebacteria [5]	eukaryotes [2]
ADP/dAMP	+	+	+
AMP/dAMP	+	+	-
Adenosine/deoxydenosine	+	+	-
Adenine/deoxyadenine	+	+	-
novobiocin	+	+	-
coumermycin A_1	+	+	-

organisms	genomic organisation
E. coli DNA gyrase [6, 7]	2 separate genes: *gyrA, gyrB*
E. coli Topo IV	2 separate genes: *gyrA, gyrB*
B. subtilis DNA gyrase [9]	2 separate genes: *gyrA, gyrB*
Haloferax DNA gyrase [10]	2 separate genes: *gyrA, gyrB* co-transcribed into one unique messenger RNA translated into two subunits
Eukaryotes DNA topoisomerase II [11]	1 gene: *topII*

Fig. 1. Structural evolution of the gene(s) coding for DNA topoisomerase II.

The NH_2 extremity of the bacterial A subunits and the similar sequence in eukarytic fused genes have little homology compared to the strong homology which is regularly found between the sequences coding for this indispensable enzyme. This variation must be useful for structuring the protein, thus leading to an efficient topography of the ATP recognition site exhibiting ATP hydrolysis potential.

The comparative analysis of various ATPases within the DNA-dependent ATPase family as well as within the DNA-independent ATPase family, gives information concerning the topography of the ATP/NTP recognition site and the potentiality for ATP/NTP hydrolysis (fig. 2). Several motifs have been already mentioned [12] which specify ATP recognition, such as the Walker sequence A/GXXGXGKT. However, we can distinguish some variations of the ATP recognition motif, between DNA topoisomerases II and DNA helicases and thus delineate several ATP-dependent enzyme families. Thus, the final lysine of the Walker sequence is absent in the case of DNA topoisomerase II. However, the lysine residue, which is located at about 10 amino acids in the front of the partial Walker sequence, has been shown to interact with ATP in *E. coli* DNA topoisomerase II [13].

Comparative analysis of several ATPases allow us to distinguish what is required at the chemical level for ATP recognition and ATP hydrolysis. A comparative analysis of all possibilities should allow us to design specific inhibitors and, furthermore, specific biosensors. Moreover, it should be design biosensors with specific affinity to either inactive or active forms of DNA topoisomerase II or any other particular enzyme.

	Walker motif
E. coli Rad3	LEMPS**G**TKG**KT**
E. coli UvrA	ITGVS**GS****GK**S
E. coli UvrD	VL**A**GA**GS****GKT**
E. coli RecA	IY**G**PESS**GKT**
E. coli RecB	IE**A**SA**G**T**GKT**
E. coli DnaA	LY**G**GT**G**L**GKT**
E. coli DnaB	VA**A**RPSM**GKT**
DNA topoisomerases II	
E. coli (103, 110) **K**, **K**	(113-119) **GG**LN**G**V**G**
B. subtilis (103, 110) , **K**	(115-121) **GG**LM**G**V**G**
Phage T4 (103, 110) , **K**	(130-136) **GG**MM**G**V**G**
S. pombe (103, 110) , **K**	(151-157) **GG**RN**G**Y**G**
S. cerevisiae (103, 110) , **K**	(140-146) **GG**RN**G**Y**G**
D. melanogaster (103, 110) , **K**	(141-147) **GG**RN**G**Y**G**

Fig. 2.Consensus sequences specifying the ATP/NTP recognition of several DNA-dependent ATPases and DNA topoisomerases II.

The first type of inhibitors to be designed are ATP-derivatives [3]. These derivatives are designed according to the fact that both ribose and deoxyribose and deoxyribose are recognized by the enzyme and only adenine is recognized. However, modifications such as the presence of one additional natrium group in the case of 8-azidoadenosine 5'-triphosphate show a reduced efficiency [3]. Pyrene-labeled ATP, which has been synthesis for the myosin ATPase [14] could be used as a model for designing biosensors for DNA topoisomerase II (fig. 3).

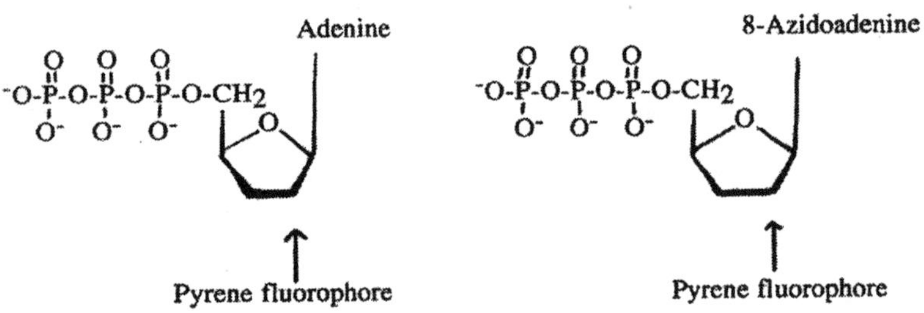

Fig. 3.ATP-derivatives.

The second type of inhibitors to be designed are coumarinderivatives [4]. The design is based on the fact that the novobiose of novobiocin and the ribose of ATP cover the same region of the ATP recognition site of *E. coli* DNA gyrase. Based on these inhibitors, specific biosensors will be designed first of all for distinguishing DNA topoisomerase II among other cellular ATPases, secondly for distinguishing either active or inactive forms of the enzyme. As the pyrene fluorophore can be linked to the ribose of ATP, a pyrene fluorophore addition to the novobiose of the novobiocin is proposed (fig. 4).

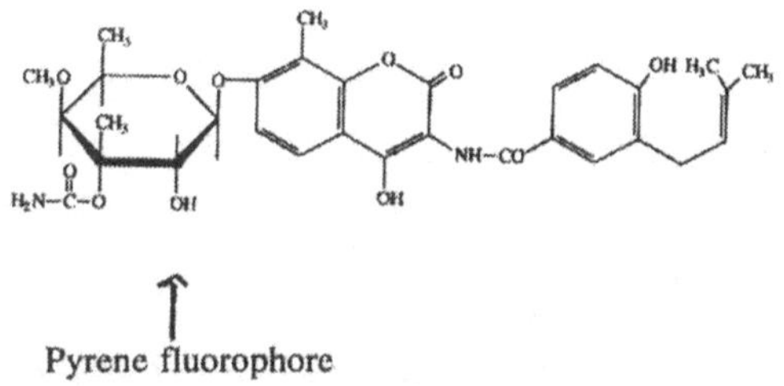

Fig. 4. coumarin-derivatives.

The use of specific biosensors should measure the variation of the activity of the enzyme which is regulated by structural changes after phosphorylation and dephosphorylation. Indeed, DNA topoisomerase II is phosphorylated during the cell cycle and the phosphorylated form of the enzyme is more active.

References

1. Wang, J., Ann. Rev. Biochem., 54 (1985) 665.
2. Assairi, L. (Ed.) Perspectives on protein Engineering. Geisow, Biodigm, 1996.
3. Assairi, L., Letters in Peptide Science, 4 (1997) 430.
4. Assairi, L., Letters in Peptide Science, 2 (1995) 169.
5. Assairi, L., Biochimica and Biophysica Acta, 1219 (1994) 107.
6. Swanberg, S. and Wang, J., J. Mol. Biol., 197 (1997) 729.
7. Adachi, T., Mizuuchi, M., Robinson, E., Appella, E., O'Dea, M., Gellert, M. and Mizuuchi, K., Nucl. Acids Res., 15 (1987) 771.
8. Kato, J., Nishimura, Y., Imamura, R., Niki, H., Hiraga, S. and Suzuki, H., Cell, 63 (1990) 393.
9. Moriya, S., Oasawa, N. and Yoshikawa, H., Nucl. Acids Res., 13 (1985) 2251.
10. Holmes, M. and Dyall-Smith, M., J. Bact., 173 (1991) 642.
11. Wycoff, E., Natalie, D., Nolan, J. M. and hsieh, T., J. Mol. Biol., 205 (1989) 1.
12. Walker, J. E., Saraste, M., Runswick, M. J. and Gay, M. J., EMBO J., 8 (1982) 945.
13. Wigley, D., Davies, G., Dodson, E., Maxwell, A. and Dodson, G., Nature, 351 (1991) 624.
14. Hiratsuka, T., Biophysical J., 72 (1997) 843.

Applications of optical biosensors to structure-function studies in the EGF/EGF receptor system

E.C. Nice[a], T. Domagala[a], F. Smyth[a], N. Konstantopoulos[a], M. Nerrie[a], D. Geleick, B. Catimel and A.W. Burgess[a]

Ludwig Institute for Cancer Research, Melbourne Tumour Biology Branch and [a]The CRC for Cellular Growth Factors, Melbourne, 3050, Australia

Introduction

Many biological signaling pathways are regulated by specific protein-protein interactions. The ability to measure such interactions in real time with high sensitivity using optical biosensors (e.g. BIAcore [1], Iasys [2]) has resulted in the rapid expansion in the use of these technologies. In these studies one of the interactants is immobilized onto a sensor surface and other reagents are applied in solution over the surface using either flow or cuvette based hydraulics. By careful control of the immobilization chemistry it is possible to attach a wide range of compounds **in an active orientation** onto the surface [3]. Detection is based on evanescent wave technology in which refractive index changes at, or near, the sensor surface modulate the signal [1]. Typically, since the detectors are mass sensitive, the smaller molecular weight compound is immobilized to maximize sensitivity of detection. Applications include protein-protein, protein-peptide, DNA-protein, DNA-DNA and lipid-protein interactions [4, 5]. Biosensor techniques have been applied for example, to, antibody-antigen, receptor-ligand, signal transduction, adhesion molecules and nuclear receptor studies.

The general applicability of the technique renders it ideally suited to structure-function studies, where it can be used to generate data or facilitate design of subsequent experiments (fig.1). In this communication we review such biosensor applications and illustrate the power and flexibility of this approach with recent examples from our laboratory.

(1) Screening : Biological Extracts
: Recombinant Production
(2) Monitoring of Protein Purification
(3) Kinetic/Thermodynamic Analysis of Binding Data
(4) Analysis of Chemical and Biological Inhibitors
(5) Detection of Novel Binding Partners (Ligand Searching)
(6) Analysis of Clinical Samples for Bioactive Components

Fig. 1.Applications of Optical Biosensors.

Results and Discussion

Screening

The first requisite of any structure-function study is the generation of suitable, well characterized reagents. The biosensor can be used to screen crude biological extracts (e.g. tissue culture medium, tissue extracts, biological fluids) for the presence of an appropriate binding partner. However, since the technique will recognize any biomolecular interaction (including non-specific or biologically irrelevant interactions [6, 7]), appropriate controls on specificity must be included. These include competition studies [7] or the use of appropriate control channels on which, for example, a denatured form of the compound of interest or a closely related, but not cross-reactive, molecule has been immobilized.

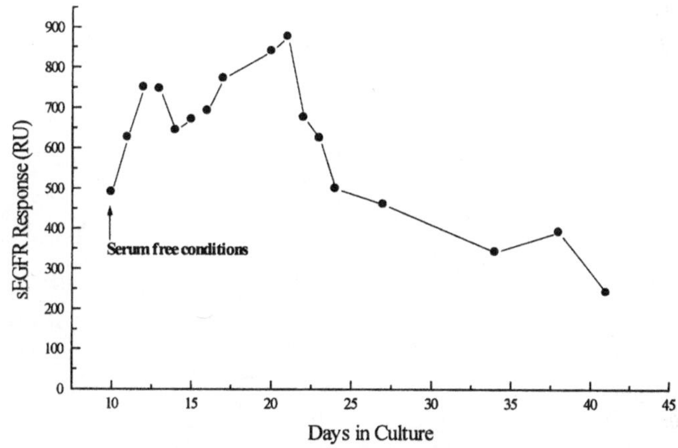

Fig. 2. Biosensor analysis of sEGFR-containing conditioned media produced in a bioreactor. Serum free production was commenced at Day 10. Aliquots (30 μl) of daily harvest were passed over immobilised rhEGF. The specific signal identified following competition with 5μg rhEGF is shown.

The use of the BIAcore to monitor the production of a recombinant form of the extracellular domain of the epidermal growth factor receptor (sEGFR) produced in a hollow fibre bioreactor is shown in Figure 2. RP-HPLC purifed rhEGF was immobilized on the sensor surface. Specificity was shown by ablation of the specific response by pre-incubating the samples (20 min. at 20 °C) with recombinant human (rh) EGF (5 μg). Signals can be quantified by reference to an appropriate standard curve (see Fig. 3).

Biosensor Compatability with Micropreparative HPLC and Mass Spectrometry.
Optical biosensors can be used in conjuction with micropreparative HPLC [8] to monitor the purification of specific proteins. Indeed, an extension of this strategy has been the basis of our ligand searching strategies [7, 9, 10]. In this case the target protein is initially purifed to homogeneity and immobilized on a sensor surface which is used for screening a biological library to identify a suitable source of binding partner. The biosensor can then be used as an affinity detector to monitor the purification. It is possible to recover samples for futher analysis from the sensor surface of the BIAcore 2000 and IAsys systems. It has been shown that sufficient material can be isolated from crude samples for direct mass spectrometric analysis on desorbed samples or the sensor chip can be used directly as a MALDI target [11].

Kinetic/Thermodynamic Analysis.

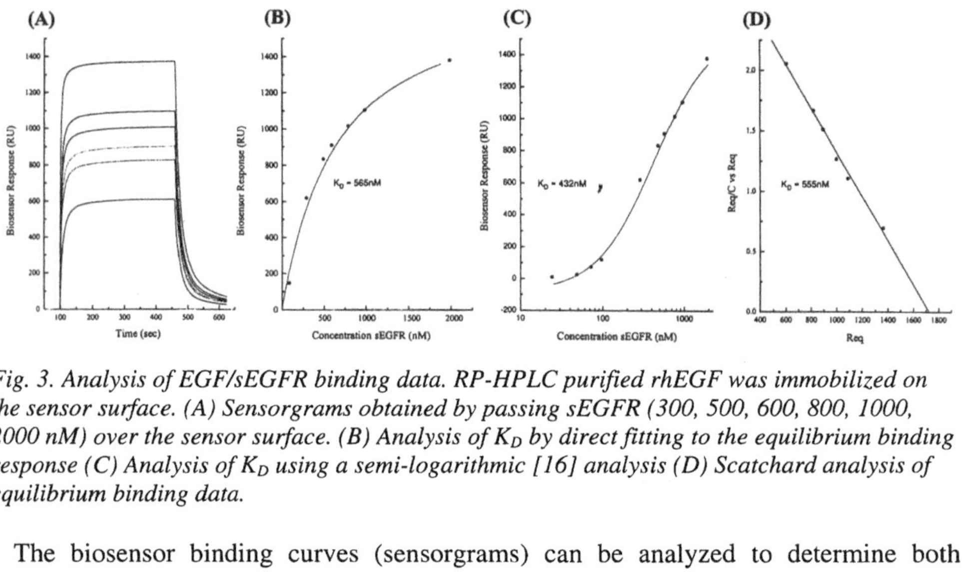

Fig. 3. Analysis of EGF/sEGFR binding data. RP-HPLC purified rhEGF was immobilized on the sensor surface. (A) Sensorgrams obtained by passing sEGFR (300, 500, 600, 800, 1000, 2000 nM) over the sensor surface. (B) Analysis of K_D by direct fitting to the equilibrium binding response (C) Analysis of K_D using a semi-logarithmic [16] analysis (D) Scatchard analysis of equilibrium binding data.

The biosensor binding curves (sensorgrams) can be analyzed to determine both equilibrium and rate constants. This can be achieved using a number of different approaches to fit the biosensor data, including linearization of the primary data [12], nonlinear least squares analysis [13] and, more recently, global fitting [14]. We have found it useful to apply a number of alternative approaches to each data set to verify the appropriateness of the models [15].

An example of quantitative analysis of the sensorgrams for the interaction between immobilized rhEGF and sEGFR is shown in Fig. 3. A series of sensorgrams was generated using different concentrations of sEGFR (fig. 3A). From the biosensor data the equilibrium binding constant can be obtained in a number of ways. In panel B, a plot of the equilibrium binding response versus the sample concentration (i.e. a standard curve) has been fitted directly to a mathematical equation describing steady state affinity using non-linear least squares analysis. In panel C the data is presented in a semi-logarithmic (Klotz) form [16] and then fitted to a sigmoidal function, while in panel D the data is presented in Scatchard format (Req/C versus Req, equivalent to B/F versus B) [17]. In all cases the data are well fitted giving a K_D of approximately 500 nM. This value was supported by non-linear least squares analysis of the individual association and dissociation phases or global analysis to determine the individual rate constants, where values of $k_d = 0.043\ s^{-1}$ and $k_a = 7.84 \times 104\ M^{-1}$ (K_D = 554 nM) were obtained for the predominant interaction.

This approach can be used to compare affinities of different batches of material in a quality control mode, or investigate the effect of specific chemical or enzymatic modifications (e.g. deglycosylation, dimerization, conjugation).

Competition Assays

As indicated above, specific binding signals can be ablated by solution competition. The biosensor can therefore be used to analyze reagents as potential inhibitors of ligand binding. The results of typical competition assays using immobilized rhEGF in competition with a number of EGF analogues binding to sEGFR in solution are shown in Fig. 4.

Fig. 4A shows the reduction in biosensor signal with increasing concentrations of rhEGF in solution. The relative potency of rhEGF, mEGF and an mEGF derivative ($EGF_{1\text{-}45}$, obtained by enzymatic cleavage of mEGF with trypsin [18]) in reducing the binding of sEGFR to immobilized rhEGF is shown in Fig. 4B. These data are in good agreement with the relative activities observed using mitogenic or cell binding assays [18]. Using automated instruments, large numbers of potential antagonists can be assayed with typical assay times of around 5 min for each sample.

The instrument can also be used in the determination of solution affinity. Thus, once a suitable calibration curve is constructed (e.g. fig. 3B), the biosensor can be used to determine directly concentrations of unbound (free) reagent in competitive binding experiments. Since the starting concentration is known, Scatchard plots can be readily constructed to determine the solution affinity without the need to assume a particular stoichiometry. Indeed the stoichiometry itself can be determined by using a modification of the Scatchard plot in which both axes are divided by the constant concentration of competitor used.In this case the intercept on the x axis gives the ratio of the interacting species [19].

Conclusions

Instrumental biosensors capable of measuring biomolecular interactions in real time are ideally suited to monitor the purification and relative binding activity of new reagents prepared for structure-function studies. The sensitivity and speed of the biosensor analysis renders it suitable for screening large numbers of samples or performing detailed kinetic analysis of the highly purified components of the interacting system.

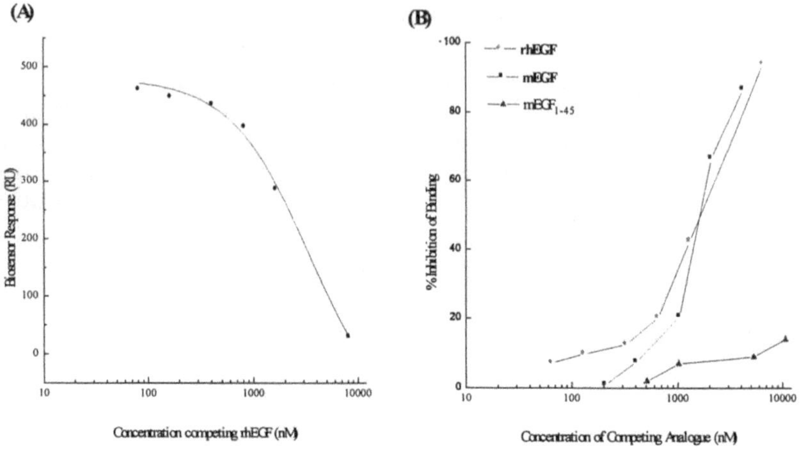

Fig. 4. Competition analysis using the BIAcore.

References

1. Malmqvist, M., Nature, 361 (1993) 186.
2. Davies, R.D., Edwards, P.R., Watts, H.J., Lowe, C.R., Buckle, P.E., Yeung, D., Kinning, T.M. and Pollard-Knight, D.V., Techniques in Protein Chemistry V, (1994) 285.
3. Johnsson, B., Lofas, S., Lindquist, G., Edstrom, A., Muller Hillgren, R.M. and Hansson, A. J., Mol. Recognit., 8 (1995) 125.
4. http://www.biacore.com
5. http://www.affinity-sensors.com
6. Cullen, D.C., Brown, R.G. and Lowe, C.R., Biosensors, 3 (1987) 211.
7. Nice, E.C., Catimel, B., Lackmann, M., Stacker, S., Runting, A., Wilks, A., Nicola, N. and Burgess, A.W., Letters in Peptide Science, 4 (1997) 107.
8. Nice, E.C., Lackmann, M., Smyth, F., Fabri, L. and Burgess, A.W., J. Chromatogr., 660 (1994) 169.
9. Catimel, B., Ritter, G., Welt, S., Old, L.J., Cohen, L., Nerrie, M.A., White, S.J., Heath, J.K., Demediuk, B., Domagala, T., Lee, F.T., Scott, A.M., Tu, G.F., Ji, H., Moritz, R.L., Simpson, R.J., Burgess, A.W. and Nice, E.C., J. Biol. Chem., 271 (1996) 25664.
10. Lackmann, M., Bucci, T., Mann, R.J., Kravets, L.A., Viney, E., Smith, F., Moritz, R.L., Carter, W., Simpson, R.J., Nicola, N.A., Mackwell, K., Nice, E.C., Wilks, A.F. and Boyd, A.W., Proc. Natl. Acad. Sci., 93 (1996) 2523.
11. Krone, J.R., Nelson, R.W., Dogruel, D., Williams, P. and Granzow, R., Anal. Biochem., 244 (1997) 124.
12. Karlsson, R., Altschuh, D. and Van Regenmortel, M.H.V., In Van Regenmortel, M.H.V (Eds.) Structure of Antigens. Volume 1, CRC Press, Boca Raton, USA, 1992, p. 127.
13. O'Shannessy, D.J., Brigham-Burke, M., Soneson, K.K., Hensley, P. and Brooks, I., Anal. Biochem., 212 (1993) 457.
14. Roden, L.D. and Myszka, D.G., Biochem. Biophys. Res. Commun., 225 (1996) 1073.
15. Nice, E.C., McInerney, T.L. and Jackson, D.J., Mol. Immunol., 33 (1996) 659.
16. Klotz, I.M., Science, 217 (1982) 1247.
17. Scatchard, G. And Ann. N.Y., Acad. Sci., 51 (1949) 660.

18. Burgess, A.W., Lloyd, C.J., Smith, S., Stanley, E., Walker, F., Fabri, L., Simpson, R.J. and Nice, E.C., Biochem., 27 (1988) 4977.
19. Zeder-Lutz, G., Wenger, R., Van Regenmortel, M.H.V. and Altschuh, D., FEBS Letters, 326 (1993) 153.

Experimental study of osteogenic growth peptide promoting bone mass in rat osteoporosis

De-Yuan Shi[a], Chao Yu[b], Tong-Yi Chen[a], Mo-Yi Li[b], Da-Fu Cui[b] and Zhong-Wei Chen[a]

[a]Shanghai Medical University, ZhongShan Hospital, Department of Orthopedics, Shanghai, 200032; [b]Shanghai Institute of Biochemistry, Academic Sinica, Shanghai, 200031, China

Introduction

Animal bone is always in metabolic balance, that is, the resorption of old bone by osteoclast and the formation of new bone by osteoblast. Under the condition of osteoporosis, bone resorption is distinctly stronger than bone formation. Therefore an effective drug for inducing osteogenesis is urgently needed for clinical purposes. Osteogenic Growth Peptide (OGP) found in bone marrow culture fluid can induce the increase of osteoblasts [1, 2]. Up to now however, no report on the phamacodynamics of the animal body as a whole has been found. In our laboratory, SD rats were fed a low alccium diet to induce osteoporosis, then were treated with OGP to observe its therapeutic effect compared with salmon calcitonin (sCT).

Method

1. Model test and drug potency

The osteoporosis model involved by feeding SD rats with a low alccium diet for 6 months. These rats were then fed normal food and were divided into 4 groups: one group was used as control (group **D**, N.S.) and the rest were treated, respectively, with OGP (group **A**, 0.2ug/0.5ml; group **B**, 2 ug/0.5ml) and sCT (group **C**, 0.1 ug/0.5 ml), by subcutaneous injection every 2 days for 2 months. The parameters assayed before and after treatment were serum AKP, BGP, bone density and the shape of bone trabecula. The instruments used for these assays were QDR-2000 DXA scanning instrument for bone density, Hitachi S-520 scanning electronic microscope and JEM-1200EX trasmission electronic microscope.

2. Synthesis of OGP

OGP was synthesized step by step from C-terminal to N-terminal according to its sequence by solid phase method. The N-terminal of each amino acid was protected by Boc group and each side chains with the appropriate protecting group. After HF cleavage and HPLC purification, an analytically pure product was obtained. Its amino acid analysis was in accordance with the theoretical value.

Result and Discusion

1. Results of bone density assay

Table 1.The value of bone density of each group (X±SD, g/cm^{2}).

group	Femur bone	Tibia bone	Lumbar vertebra
A (n=7)	0.2548±0.0177	0.2623±0.0234	0.2434±0.0270
B (n=7)	0.2742±0.0264	0.2649±0.0258	0.2816±0.0667
C (n=7)	0.2635±0.0517	0.2997±0.0416	0.2675±0.0320
D (n=7)	0.1741±0.0241	0.1686±0.0177	0.1880±0.0093

Table 1 shows that the bone density of OGP group (A, B) and sCT group (C) is distinctly higher than that of control group (D).

2 Results of serum AKP assay

Table 2.Result of serum AKP assay of each group (X±SD, U/L).

Group	A (n=10)	B (n=10)	C (n=10)	D (n=10)
	169.7±16.4	165.8±11.7	100.8±9.3	83.4±9.5

Table 2 shows that AKP value (or level) of OGP group (A, B) is much higher than that of the control group D ($P<0.001$) and also higher than sCT group C ($P<0.01$).

3. Results of BGP assay

Table 3. Result of serum BGP assay of each group (X±SD, ng/ml).

Group	A (n=10)	B (n=10)	C (n=10)	D (n=10)
	2.17±0.20	2.02±0.19	1.43±0.15	1.21±0.09

Table 3 shows that the BGP level of OGP therapy groups (A, B) is significantly higher than that of control group D ($P<0.05$), but there was no distinct difference between as the sCT therapy group C & control.

4 Results of scan elecronic microscope observation

The bone trabecula of the control group was thin, broken and discontinuous, but the bone trabecula of the therapy groups (A, B) was thick, continuous and intact.

5 Results of transmission electronic microscope observation.

*TE*M showed that the bone cells of the control group showed degenerative changes with pyknotic nuclei and decreased cytoplasm, but the bone cells of therapy groups were in a formative osteogenic phase with narrow lacunar gaps and rich in cytoplasm..

Conclusion

OGP can simulate osteogenic activity, raise the level of serum AKP and BGP, enhance new bone formation and inhibit bone resorption. It thus has a potential role in treating osteoporosis and fracture.

References

1. Bab, I., Gazit, D. and Muhlrad, A., Endocrinology, 123 (1988) 345.
2. Greenberg, Z., Chorev, M. and Muhlrad, A., Journal of Clinical Endocrinology and Meyabolism, 80 (1995) 2330.

Determination of inhibitory constants for CPA by competitive spectrophotometry

Nam-Joo Hong, Jeoung-Wook Lee, Seong-Hun Chang and Dong-Hoon Jin

Yeungnam University, Gyungsan City, Gyungbuk, 712-749, Korea

Introduction

Competitive spectrophotometry [1-3] is known as a powerful tool for determining the kinetic constants of enzyme catalyzed reactions. This method relies on measuring the progress curve for a detector substrate (whose *K*m and *V*max are known) in the presence of a competing substrate. Although only the cleavage of the detector is monitored, the presence of the competing substrate alters the progress curve in a systematic manner that depends on the ratio of their kinetic constants. Since the ratio of kinetic constants is of much importance, competitive spectrophotometry represents a convenient way of measuring the kinetic constants of substrates whose rates can not be measured by progress curve analysis.

In order to find alternative substrates with more powerful kinetic constants, we have measured the kinetic constants of various natural analogs containing different N-blocking groups using Ac-Phe-S-Phe as a detector by competitive spectrophotometry.

Results and Discussion

The progress curve for an enzyme obeying Michaelis-Menten kinetics in the presence of a product inhibitor is:

$$t = y/Va + (Ka/Va)(1+(Bo/Kp))\ln(So/So\text{-}y) + (1-(Kb/Kp))(Bo/Vb)[1-(So\text{-}y/So)^n].$$

When there is an alternative substrate present, the equation is modified to:

$$t` = y/Va + (Ka/Va)(1 + (Bo/Kp))\ln(So/So\text{-}y).$$

This is possible since all the constants are known; a series of y values simply needs be inserted to calculate the corresponding t values.

The construction of the theoretical curve allows the following:

$$t` - t = Dt = (1- (Kb/Kp))(Bo/Vb)[1-(So\text{-}y/So)^n].$$

Rearranging:

$$(So\text{-}y/So)^n = (Vb/Bo)/(1- (Kb/Kp))\, Dt + 1.$$

Thus the value of n may be obtained in the same way as when only the two competing substrates are present. The y-intercept of this plot is still at 1, and the x-intercept of the plot is equal to:

$$(1-(Kb/Kp))(Bo/Vb) \text{ and } n = (Ka)(Vb)/(Kb)(Va).$$

Thus, there are two equations with two unknowns, and *Kb* and *Vb* may be determined.

Table 1. n values, Km, and Vmax of Succ-PP, Chac-PP, Cbz-PP and Ac-PP with Kp value from Dixon plot(CPA.)

Substrate	*Kp* (mM)		n		*Km* (mM)		*Vmax* (ΔOD /min)	
	(Dixon)	(cal)*	(Dixon)	(cal)	(Dixon)	(cal)	(Dixon)	(cal)
Succ-PP	.86	.44	8.87	11.10	.041	.038	1.42	1.63
Ac-pp	.86	.64	10.18	12.10	.069	.066	2.74	3.08
Cbz-pp	.86	.35	14.08	12.45	.062	.064	3.45	3.14
Chac-pp	.86	.28	13.11	17.42	.033	.031	1.68	2.09
Boc-pp	.86	.39	13.21	12.13	.058	.056	2.86	2.72

* *Calculated by computer. Note that theses values are averages of 2-4 runs.*

Since it was thought that N-blocked phenylalanine as well as free phenylalanine may inhibit CPA, an alternate method of caculating *Kp* was formulated. This method relied on the assumption that the term: $[1-(So-y/So)^n]$ approached the value of 1 as y became greater than 1/2 of So. This was assumed since most n values were expected to be equal to or greater than 6. If this assumption was correct, then Dt would also reach a constant limit of: (1-(*Kb*/*Kp*))(Bo/*Vb*). With the help of a computer, different values of *Kp* were used to make the theoretical curve that would give the most constant Dt at y values greater than 1/2 of So. The *Kp* value that gave the most constant Dt was considered the correct value. In calculating n, *Kb* and *Vmax* for the varous substrates, both values of *Kp* were used.

In regard to kinetic constants of the various N-blocked Phe-Phe compounds, no clear pattern was shown. It was speculated that substrates having more electronegative blocking groups were split faster. However, Ac-PP is split considerably faster than Chac-PP. Although the *Km* value of Ac-PP is improved somewhat by changing the N-blocking groups, there does not seem to be any significant improvement in the rate of splitting. It must not be expected that there would be any significant improvement of the kinetic constants of Ac-PSP by simply changing the N-blocking groups. There is no doubt that L-phenylalanine inhibits carboxypeptidase A. The calculation of *Kp* through the indirect method of Dt's seems to indicate that both the free and N-blocked phenylalanine residues inhibit the enzyme. *Kp* calculated from the computer is less than *Kp* calculated from the Dixon plot in all cases. Moreover, it seems that the more polar blocking groups have the greatest inhibitory effect. However, using one *Kp* value (Dixon plot) over the other (calculated) does not radically change the values for the kinetic constants. In conclusion, competitive spectrophotometry has shown that the kinetic constants of Ac-PP were not improved by changing the N-blocking group.

References

1. Hong, N.J., Bull. of Korean Chem. Soc., 19 (1998) 189.
2. Hong, N.J., Bull. of Korean Chem. Soc., 19 (1998) 687.
3. Hwang, S.E., Brown, K. S. and Glivarg, C., J. Med. Chem., 170 (1988) 161.

Sublingual administration of monomeric insulin -- destetrapeptide insulin

Mo-Yi Li[a], Hong Peng[b], Min-Rong Ai[a], Hui-Bi Xu[b] and You-Shang Zhang[a]

[a]*Shanghai Institute of Biochemistry, Academia Sinica, Shanghai, 200031;* [b]*School of Science, Huazhong University of Science and Technology, Wuhan, 430074, China*

Introduction

Insulin has been used clinically for over 70 years. The usual way of administering insulin is by subcutaneous injection in the form of hexamers which have to be dissociated *in vivo* to form monomers before insulin can perform its biological function [1-3]. In order to avoid the time lag caused by dissociation, monomeric insulin has been developed for clinical application [4]. In addition, oral administration of insulin is more convenient than injection but the resulting bioavailability is not sufficiently high. Destetrapeptide insulin (DTI), with the last four residues of the insulin B-chain removed, has been found to be monomeric (data not shown). In contrast to monomeric insulin reported previously, the amino acid sequence is the same as insulin except for the shortened B-chain. Here, we report the enzymatic semisynthesis of DTI and its hypoglycemic effect in rats by sublingual administration. Because DTI molecules are smaller and do not associate, it is better absorbed. Therefore, the bioavailability of DTI is higher than that of insulin.

Experiments and Results

Preparation of DTI

Porcine insulin (3 mg) was dissolved in 0.2 ml of 0.2 M Tris containing 85% 1,4-butanediol. GFFY acetate salt in excess (10:1 in molar ratio) was added and the pH of the reaction mixture was adjusted to 8.0-8.5 by N-methylmorpholine. Trypsin (0.6 mg) was added and the reaction mixture was incubated at 30°C for 24 hours. To stop the reaction, 0.08 ml of acetic acid was added. The crude product was purified by Sephadex G-50 column and HPLC C8 column chromatography and 0.52 mg of the purified product was obtained. It was homogeneous in PAGE and capillary electrophoresis (CE) (fig. 1 and 2). The molecular weight was confirmed by mass spectroscopic analysis (fig. 3). The overall yield was 17%. The biological activity of DTI was determined as 22 IU/mg by mouse convulsion method according to British Pharmacopoeia.

Hypoglycemic effect of DTI in rats by sublingual administration

DTI or insulin was mixed with Lp101 promoter and administered sublingually to SD rats [5]. The hypoglycemic effect (D%) was calculated according to the following equation:

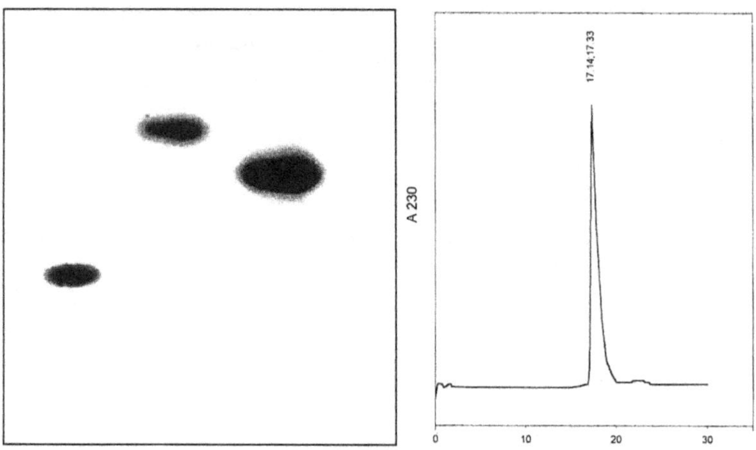
17.14;17.33
A 230
0
10
20
30

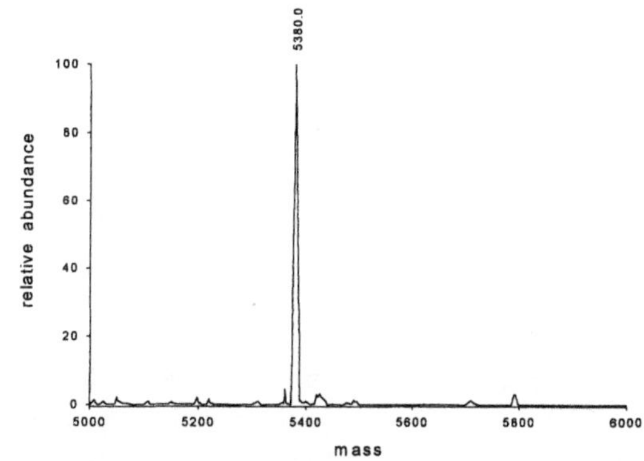

Fig. 3. Mass spectroscopic analysis of DTI MW calculated 5380, determined 5380.

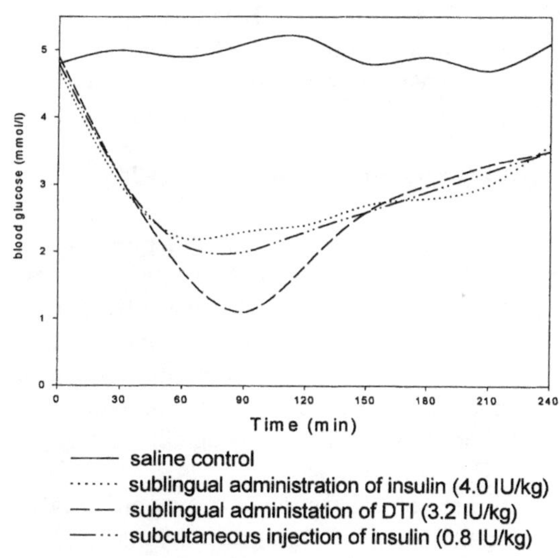

Fig. 4 Hypoglycemic effect of DTI and insulin.

References

1. Blundall, T., Adv. Protein Chem., 26 (1972) 279.
2. Drejer, K., Diabetes/Metabolism Reviews, 8 (1992) 259.
3. Insulin Research Group, Academia Sinica, Sci. Sin., 17 (1974) 779.
4. Brange, J., Nature, 333 (1988) 679.
5. Dondeti, P., Int. J. Pharm., 122 (1995) 91.

How to get patent rights to protect your invention in biotechnology

Li Zhou
Chemical Examination Department, Chinese Patent Office Beijing, Beijing, 100088, China

Introduction

During the past ten years, biotechnology has been developing rapidly. The progress of biotechnology has greatly accelerated the development of the economy. In the meantime the competition in this field has become more and more fierce, and patent protection of these new techniques has played an important role. It is believed that the 21st century will be the biotechnological century. If you want to be the winner in the next century, you should obtain a patent to protect your new invention. Not only can this protect you, but it also can prevent you from infringing other people's rights. The patent rights will be your shields against attack in the competitive research area.

What can be filed for patent in biotechnology?

As all we know, a patent can be applied to a product, a process and a kind of use. In biotechnology, the products which can be patented include chemical products, for example, antibiotics, peptides, genes, pharmaceutical compositions; living products, i.e. cellular organisms, such as bacteria, fungi, animal and plant cells and lines, non cellular organisms such as viruses; and apparatus or kits. The processes which are patentable are processes for preparing these products, for example, a process for producing a product by fermentation. The uses which are patentable are uses of products, such as medical use.

However, there are some exceptions to patentability in biotechnology: Discoveries, methods of treatment of humans or animals by surgery, therapy and diagnostic methods practiced on the human or animal body, plant and animal varieties or essentially biological processes for the production of plants or animals [1]. Thus gene therapy can not be patented. Recently with the development of the human genome study, there are many applications about DNA or RNA fragments without any functions. These fragments whose functions are not known can not be patented either.

When should the application be filed? What are the tactics?

Because novelty is the most essential requisite of an invention, you must make a novelty search before you prepare to file an application for patent.

According to Chinese Patent Law, the person who first files the application will get the patent right. Thus if you want to publish your newest invention in a journal or give an oral presentation at a conference, it's better for you to file patent application first, even though Chinese Patent Law stipulates an article on disclosures not causing loss of novelty.

Use priority right as far as possible. Article 29 of Chinese Patent Law says, "Where, within twelve months from the date on which any applicant first filed in a foreign country an application for a patent for invention, he or it files in China an application for a patent for the same subject matter, he or it may, in accordance with any agreement concluded between the said foreign country and China, or in accordance with any international treaty to which both countries are party, or on the basis of the principle of mutual recognition of the right of priority, enjoy a right of priority. Where, within twelve months from the date on which any applicant first filed in China an application for a patent for invention..., he or it files with the Patent Office an application for a patent for the same subject matter, he or it may enjoy a right of priority." Therefore, first file a priority application. You can then continue your study and improve it later. Within 12 months file another document based on the prior one but more perfect. In this way you can enjoy the priority date while the document substantively examined by the patent office is the latter one.

On Jan. 1st, 1994 Chinese Patent Office (abbreviated as CPO) joined the Patent Corporation Treaty (abbreviated as PCT). From then on CPO has become an international receiving office, search office and preliminary examination office. If you want to file one application in many countries, you can use the PCT procedure in CPO. However, as it's expensive and complex, it's not suitable for small companies or personal applicants [2].

How to write the application in biotechnology?

Description and claims constitute the two main sections of an application for patent. They must meet the requirements of article 26 of Chinese Patent Law and its rules 18 to 23 and 25.

The description shall set forth the invention in a manner sufficiently clear and complete so as to enable a person skilled in the relevant field of technology to carry it out. Inventions of biotechnology often involve microorganisms. If the micro-organisms are obtained by mutation or isolated from natural sources, they will not be available to the public. In this case you must deposit it in a depository institution which is recognized by CPO, such as CCTCC and CGMCC in China.

Claims must be clear and concise and be supported by the description. Here I'd like to give some examples on how to write them in biotechnology. With respect to product claims, for example a protein, you may use three forms. First, describe the protein by its sequence structure. Another form is to describe the protein in parameters, i.e. physical and chemical parameters, such as molecular weight, isoelectric point etc.. The third method is to characterize the protein as a product by a process. For a new microorganism, there are three possible ways to write a claim. First, define the microorganism in a claim by the depository number and the name of the institution. The name of the genus and species to which the strain belongs should also be quoted, although this is not obligatory. Thus, such a claim could read, "Microorganism of the (genes) (species) (depository institution) (the depository number)." Second, the microorganism can be described by the process by which it is obtained provided that the process used is reproducible. Third, the microorganism can be described by means of parameters. As for medical use claims, since therapeutic and diagnostic methods can not be patented, you can claim the use of a compound which is pharmaceutically active in this way "The use of compound A for preparing the pharmaceuticals for treatment of X disease."

References

1. Guidline of Chinese patent examination, Section 2, Chart 1, Periods 3.3 and 3.4
2. Wei, S., Intellectual Property, 3 (1998) 29.

Study on peptide maps from milk protein hydrolysates by SE-HPLC

Li Zhao, Yan-Qun Li, Zi-Lin Chen, Shi-Wang She and Yong-Fu Wang
Jiangxi-OAI joint research institute, Nanchang, 330047, China

Introdution

Because milk protein hydrolysates have been shown to possess various physiologically active peptides, with immunomodulation, amtithrombotic and antihypertensive majesties, they have been the subject of much research. In our lab, a new functional food milk peptide powder whose peptides range in molecuar weight from 1000 D to 8000 D has been prepared from milk protein hydrolysates. In these paper, a series of peptide maps under different hydrolysis conditions (pH, Temp., Time, Category of enzyme) were researched by size exclusion-HPLC.

Materials and Methods

Fresh milk was purchased from local market, and enzyme I, II, III were purchased from Biochemical Reagent Comp., Shanghai. The milk peptide powder was produced by enzyme hydrolysis (fig. 1).

Mixed enzyme
↓
Fresh milk--→Skim milk---→Hydrolysate--→Inactivation of enzymes----→Removing the bitterness--→Enrichment--→Spray drying--→The milk peptide powder.

Fig. 1.Processing route of the milk peptide powder.

The molecuar weight range of peptides in the hydrolysate was examinated by size exclusion-HPLC. Size exclusion-HPLC was carried out on a Protein-PAK ™ 60 column (300 × 7.8 mm I.D.) using a HPLC system equipped with a Model 510 pump, a universal liquid chromatography injector with a 100 loop, a Lambda Max Model 481 LC spectrophotometer (Milipore) and a Baseline 810 chromatography workstation. The mobile phase was 0.05 M pH7.1 phosphate buffer. The system was run isocratically at a flow rate 0.5 ml/min at constant temperature. Polypeptides were monitored at 214 nm with an absorbance scale 0.05. The mobile phase was filtered through a 0.45 μm filter (Milipore) and degassed by ultrasonication before use. The standard molecular weight markers for peptides, 2512-16949 D, were purchased from Pharmacia. In this test, an even design was used, and five factors were evaluated: T=time of enzyme hydrolysis, Temp.= temperature of the enzyme hydrolysis, enzyme I, II, III=three kinds of enzymes that used in the hydrolysis.

Results and Discussion

The retention time (HPLC) of the standard molecular weight markers for peptides was determined first (table 1). Then, twelve tests in even design were performed (table 2). The profiles of the standard molecular weight markers for the peptide map (fig. 2) and the milk peptide map were obtained from size exclusion-HPLC. Table 1 shows that, when retention time was in the 16.00-26.00 min range, the distribution of peptide molecular weights ranged from about 2000-8000 D.

Table 1. The retention time of the standard molecular weight markers.

Mw (Dal.)	Retention time (min)
2512	21.79
6214	19.93
8159	17.77
10700	16.03
14404	11.73
16949	10.85

Table 2. The table of even design.

No	T (min)	Temp. (°C)	Enzyme I (×103 IU/g)	Enzyme II (×103 IU/g)	Enzyme III (×103 IU/g)	Peak area percent*
1	30	44	2.0	4.5	6.0	86.89
2	45	50	4.0	2.5	5.5	88.81
3	60	56	6.0	0.5	5.0	86.96
4	75	62	1.5	5.0	4.5	91.66
5	90	42	3.5	3.0	4.0	89.08
6	105	48	5.5	1.0	3.5	96.14
7	120	54	1.0	5.5	3.0	90.57
8	145	60	3.0	3.5	2.5	89.02
9	170	40	5.0	1.5	2.0	90.00
10	195	46	0.5	6.0	1.5	74.48
11	220	52	2.5	4.0	1.0	78.59
12	245	58	4.5	2.0	0.5	13.31

 # Peak area percent is the area that the peak located in the retention time 16.00-26.00 min in the milk peptide map.

Regression analysis revealed the optimal results of the even design. To produce the milk peptide powder, the optimal time is 120 min, the optimal temper-ature is 40 °C and the optimal active unit of enzyme I, II, III was 5.4×10^3, 5.8×10^3, 5.8×10^3 IU/g. The milk peptide map under optimal conditions was obtained from size exclusion-HPLC (fig. 3). The distribution of the peptide molecular weights of the milk peptide powder was mostly in the range of 2000-8000 D.

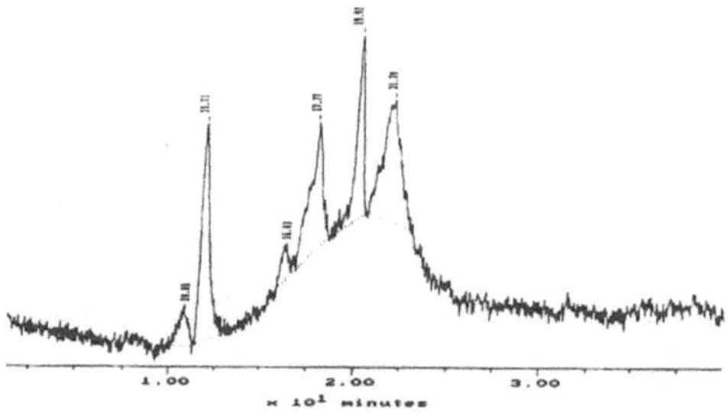

Fig. 2. The profile of the standard molecular weight markers for peptides.

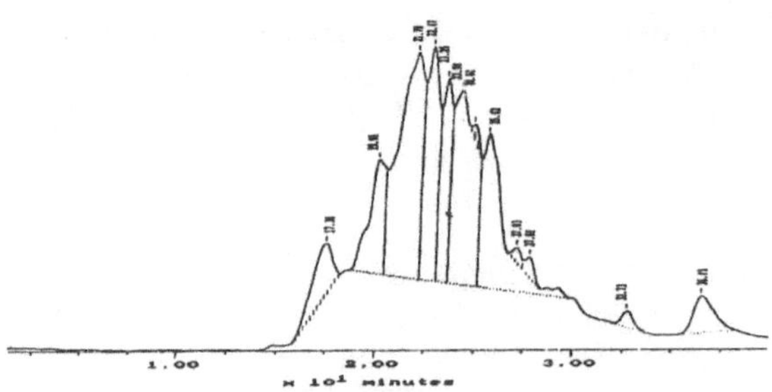

Fig. 3. The profile of the milk peptide map.

Acknowledgments

This work was supported by grants from Natural Science Foundation of Jiangxi (No.95307).

References

1. Siemensma, A.D. and Weijer, W. J., Trends in Food Science and Technology, 4 (1993) 16.
2. Chen, Z.L., Zhao, L., Li, Y. Q. and She, S.W., Food and Fermention in Hunan, 1 (1997) 32.
3. Zhao, L., Li, Y.Q. and She, S.W., Food Research and Development, 4 (1997) 8.

Synthesis and solution structures of new chiral enantiomeric peptide nucleic acid (PNA) dimers

Li-Gang Zhang, Ji-Mei Min and Li-He Zhang
National Key laboratory of Natural and Biomimetic Drugs, Beijing Medical University, Beijing, 100083, China

Introduction

The quest to develop new drug therapies based on sequence specific interactions between complementary nucleic acids is an exciting and rapidly growing field of chemical research. In the early 1990s', Nielsen's group prepared the first (peptide nucleic acid) PNA oligomer [1], which has an achiral, uncharged pseudopeptide backbone consisting of N-(2-aminoethyl) glycine units while the nucleobases are attached to the glycine nitrogens via methylene carbonyl linkers. Since then, many modified PNAs have been synthesized.

In most cases, PNAs were constructed with lysine at the C-terminal of the oligomers, which was designed to improve the solubility of the whole molecule or avoid self-aggregation. Since lysine is a natural amino acid and readily available commercially, we wanted to synthesize a new type of PNA molecule, with lysine as the main chain unit and the nucleobases attached through carbonyl methylene group to the α-NH_2 of the amino acids (fig. 1). Although such a structure will separate the adjacent bases by seven bonds and the distances between the bases and the main chain would be three bonds, both one bond more than the natural DNA and PNA designed by Nielsen, we thought that the long aliphatic lysine side chain might make the molecule more flexible. It could thus accommodate itself to a suitable spatial distance when hybridizing to complementary oligonucleotides. Also, we expected the four methylene groups in the backbone unit to give the whole oligomer some hydrophobicity, which, together with the hydrophilic characteristics of lysine itself, may facilate cellular uptake. Although only the D-lysine series (Ib) are supposed to mimic natural oligonucleotides in regard to configuration, the more readily available and inexpensive L-lysine series (Ia) was also investigated.

In order to check whether the designed oligomer has the ability to adopt a less extended conformation when hybridizing, we synthesized the corresponding dimers and studied the solution conformations based on the results of 2D-NMR (NOESY) and molecular simulation.

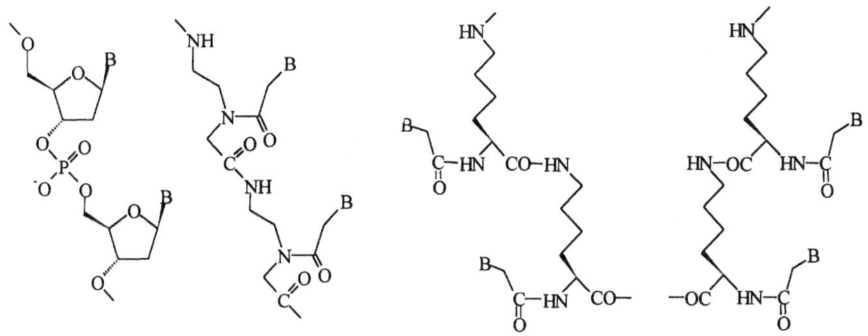

DNA PNA Ia Ib

Fig. 1. Structures of DNA, PNA[1], and PNA molecules based on L- (Ia) and D- lysine (Ib).

Results and discussion

Starting from Fmoc-L-Lys(Boc)-OH, compound 3a was prepared in two steps. The carboxylic acid was first converted to the corresponding benzyl ester with benzyl bromide and triethyl amine, and then Fmoc selectively removed by 50% Et_3N/DMF stirring at r.t. for about 10 hours with higher than 90% yield (scheme 2). The enantiomeric compound 3b was prepared using exactly the same procedures starting from Fmoc-D-Lys(Boc)-OH.

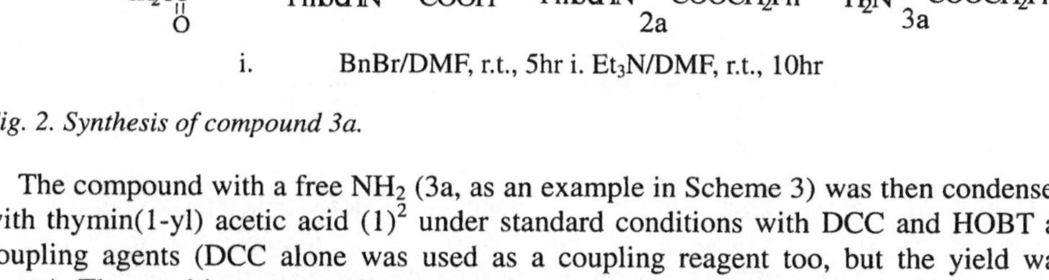

i. BnBr/DMF, r.t., 5hr i. Et_3N/DMF, r.t., 10hr

Fig. 2. Synthesis of compound 3a.

The compound with a free NH_2 (3a, as an example in Scheme 3) was then condensed with thymin(1-yl) acetic acid (1)[2] under standard conditions with DCC and HOBT as coupling agents (DCC alone was used as a coupling reagent too, but the yield was lower). The resulting protected monomer 4a can be deprotected on either side to afford compounds 6a or 5a, respectively. They can both be used directly to produce a protected dimer 7a of L-configuration (fig. 3). Similarly, the D-isomer was prepared by the condensation of compounds 6b and 5b, which, respectively, are enantiomers of 6a and 5a.

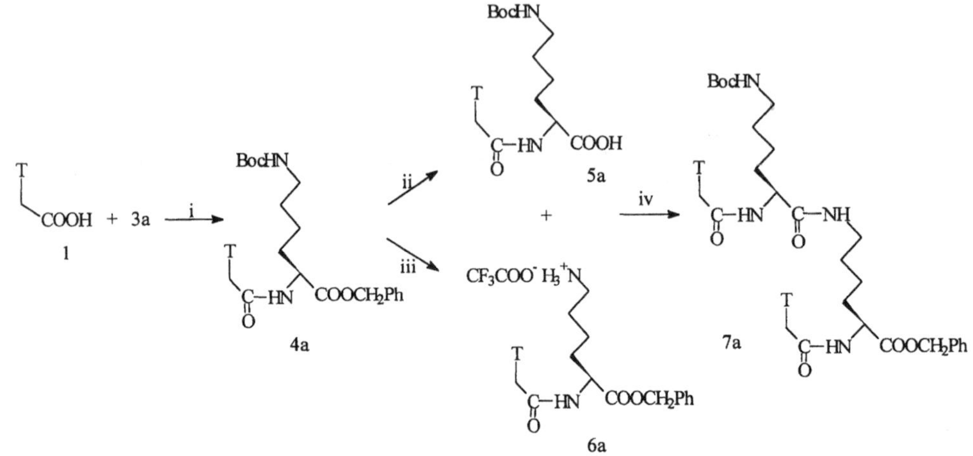

i. DCC, HOBT/DMF, 6hrs ii. H_2, 5% Pd/C, ethanol iii. 33% TFA/DCM, r.t. 5min
iv. DCC, HOBT/DMF, 12hrs

Fig. 3. Procedures for the synthesis of protected dimer 7a.

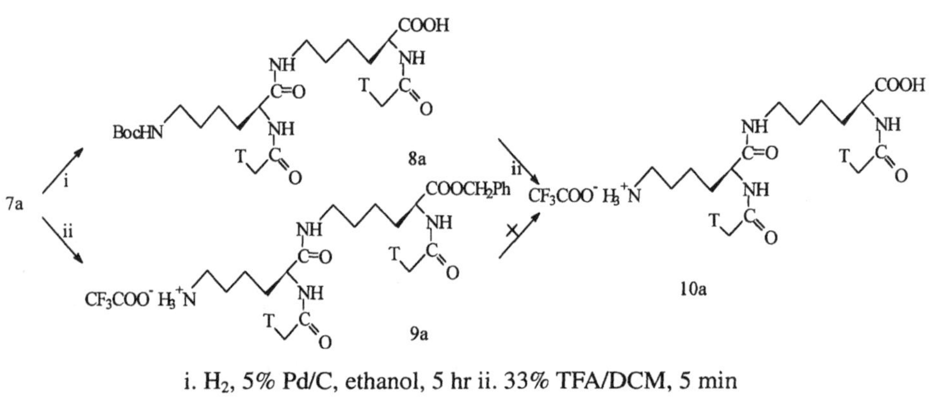

i. H_2, 5% Pd/C, ethanol, 5 hr ii. 33% TFA/DCM, 5 min

Fig. 4. Preparation of dimer 10a.

Deprotection of the benzyl ester of compound 9a did not work. Thus, dimer 10a was obtained by deprotection of 7a first through hydrogenation in ethanol for about 5 hours and then, after filtering off Pd/C, direct acidolysis to remove Boc (fig. 4).

Dimer 10b starting from suitably protected D-lysine was prepared similarly in a one spot procedure from 7b via hydrogenation and acidolysis. The crude products thus obtained were purified on HPLC with a H_2O/CH_3CN (0.1% TFA) gradient as eluent.

In order to study the solution structures of the dimers, we obtained NOESY of compound 10b in DMSO-d_6, from which the following information can be derived: the α-N<u>H</u> in the N-terminal residue has NOE correlations with hydrogens in the methylene carbonyl linkers of both residues, while the α-N$\underline{H}$ in the C-terminal residue has NOE correlation only with hydrogens in one of the two methylene carbonyl linkers; one of the two pairs of methylene hydrogens has NOE correlation with 6-H of the nucleic base(s); the N$\underline{H}$ of the peptide bond has NOE correlation with one of α-C$\underline{H}$. Putting all these data into a computer simulation program (Insight II/Discover) to constrain the distances between the corresponding atomic pairs within 3 angstroms, the results of minimization (fig. 5) indicated that the two bases as well as the two linkers are not in a parallel relationship or in the same orientation. Instead, it seems that the side chain of the C-terminal residue is close in space to that of the N-terminal residue and somewhat inserted between the base and the linker of the N-terminal residue. It thus can be inferred that the bases in such a backbone could adopt a suitable spatial conformation that is not so extended, i.e., the oligomer may have the potential to fit to the desired distances between bases when hybridizing.

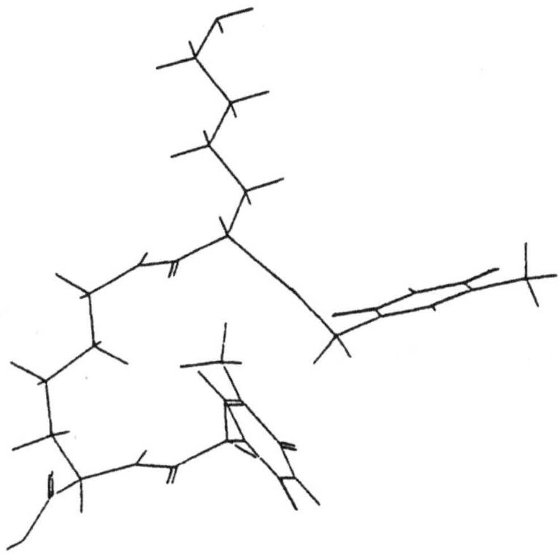

Fig. 5 The conformation of 10b simulated by Discover (N terminal: upside).

References

1. Nielsen, P.E., Egholm, M. and Berg, R.H., Science, 254 (1991) 1497.
2. Jones, A.S., Lewis, P. and Withers, S. F., Tetrahedron, 29 (1973) 2293.

Study of one type of metallothionein and SZ51 recombinant protein directing and imaging reagent

Yan Sun and Bing-Gen Ru
National Laboratory of Protein Engineering, College of Life Sciences, Peking University, Beijing, 100871, China

Introduction

Metallothionein (MT) is a kind of low molecular weight, cystine-rich, metal binding protein, widely distributed in nature ranging from bacteria to vertebrates [1]. One of the important roles of MT derives from its high metal binding capacity. Because it can totally bind 7 equivalents of bivalent metal ions, metallothion is usually used in the radioimaging and clinical detection [2].

Thrombosis is a commonly encounted clinical disorder. It occurs quite frequently in conditions such as acute myocardial infarction (AMI), pulmonary thromboembolism and deep artery or vein thrombosis. Until now the clinical diagnosis of thrombus has not been perfect [3].

Using mouse metallothionein class I (mMT-I) and SZ51, a type of monoclonal antibody which reacts specifically with α-granule membrane proteinGMP140 [4] on the surface of activated platelets during thrombus formation, a novel genetically engineezed recombinant protein was cloned and successfully expressed in the *E. Coli* BL21 (DE3) pLysS strain (Promega).

Material and Methods

We use the constructed plasmid pHENI-SZ51Fab/Hu composed of humanized SZ51 chimeric Fab fragment and plasmid pBX-MT (containing mMT-IcDNA) as template. Considering the big molecule of antibody SZ51, we used a single chain antibody of SZ51 in the form of V_H-Linker1 -V_L, and cloned with mMT-I cDNA gene (V_H-Linker1-V_L-linker2-mMT-IcDNA). Primers were synthesized and polymerase chain reactions (PCR) were processed. After that, the amplified products of the predicted size (V_H 354 bp, V_L 346 bp, mMT-IcDNA 198 bp) were verified on 8% PAG. During the PCR reactions, the suitable endonuclease digestion sites and the linkers among V_H V_L, and mMT-I cDNA fragments were designed. These PCR amplified fragments were linked and then inserted into pET5a express system (Promega) on its non-fusion protein expression site (NdeI site). A new recombinant plasmid pET5a-ScFv-mMT is shown as Fig.1.

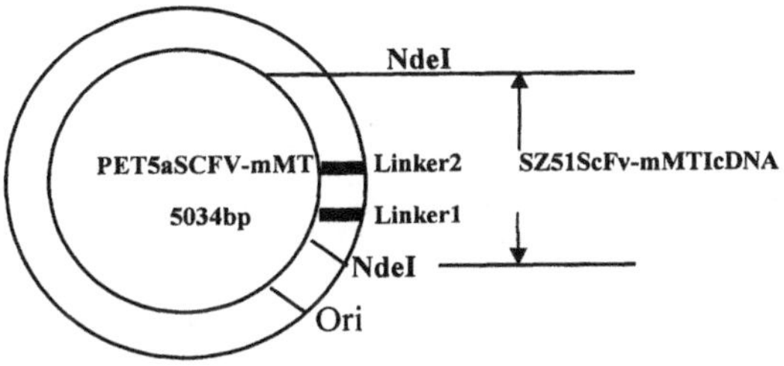

Fig. 1. Construction of recombinant plasmid pET5a-SCFV-MT.

Results and Discussion

E.Coli BL21 (DE3) pLysS cells transformed with pET5a-SCFV-MT plasmids were cultured and the non-fusion proteins were expressed after isopropyl-β-D-thiogalactoside (IPTG) was induced. The recombinant protein existed in the form of an inclusion body in total bacteria proteins. After sonication, the centrifuged inclusion body was purified by DEAE Sepharose Fast Flow ion exchange chromatography and gel filtration on Sepharose G50. Results showed that the molecular weight of the recombinant protein is 31kilodalton. The atomic absorption spectrophotometry proved that the recombinant protein had higher metal binding capacity and the enzyme-linked immunosorbent assay showed the recombinant protein not only had the same activity as the original antibody SZ51, but also native mouse metallothionein. This study will provide a kind of new, stable, effective directing and imaging reagent in the field of thrombus detection.

Acknowledgements

This work was supported by the National Ninth-five Year Key Technology Program, and the plasmid pHENI-SZ51Fab/Hu was kindly provided by Prof. Chang-Geng Ruan, Institute of Thrombosis, Suzhou Medical College.

References

1. Kagi, J.H.R. and Schaffer, A., Biochemistry, 27 (1988) 8509.
2. Wu, J., He, G., Wu, G. and Ruan, C., Nuclear Medicine Communications, 14 (1993) 1088.
3. Wang, Zh., Li, J. and Ruan, C., Thrombosis and Homeostasis, (1996) 372.
4. Gu, J., Zhang, X. and Xia, L., Chinese Journal of Hemotology, 16 (1995) 459.

LHRH antagonists: new preclinical and clinical results

B. Kutscher, M. Bernd, W. Deger, T. Reissmann, R. Deghenghi and J. Engel

Corporate research & Development ASTA Medica AG, Dresden, Germany

Introduction

Gonadorelin (GnHR, LHRH) analogues are indispensable drugs for the clinical treatment of sex-hormone dependent tumors, such as breast, ovary and prostate cancers. LHRH agonists and antagonists have also been utilized in various artificial reproduction techniques and have been investigated as potential contraceptives in humans.

Over 25 years, thousands of analogues of LHRH, both agonists and antagonists, have been synthesized and evaluated for potential therapeutic benefits. LHRH agonists, such as Buserelin and Leuprorelin, initially stimulate (7-14 days) the release of LH and FSH prior to downregulation of LHRH receptor, transiently causing rising levels of the sex steroid hormone testosterone in cancer patients suffering from metastasis related bone pain in addition to loss of libido.

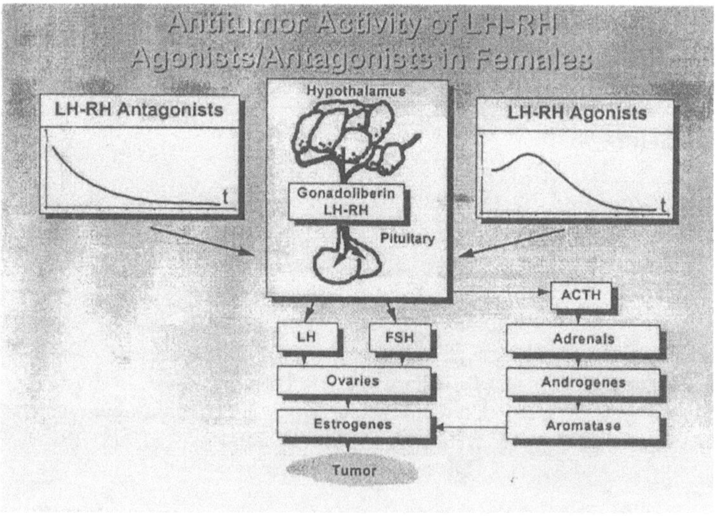

Fig. 1.

"Third generation" antagonists such as Cetrorelix are under investigation in clinical studies. Cetrorelix itself has been characterized as a potent LHRH antagonist. It does not have effects due to histamine release and induction of edema. In addition, in vivo studies confirmed its safety. Based on results in a variety of models, the potency of

Cetrorelix was clearly demonstrated. The clinical studies indicated that Cetrorelix is a safe and highly effective gonadotropin and, subsequently, sex-steroid-suppressing agent. Therefore, Cetrorelix has the potential to be used for the treatment of hormone dependent cancers as well as for non-malignant indications in which a suppression or control of gonadotropins and sex-steroids is desired.

The advantage of Cetrorelix is that it inhibits LH and testosterone from the start of administration and thereby avoids the agonist-induced "flare up effect". Therefore, the treatment of prostatic carcinoma and premenopausal mammary carcinoma, and primary induction in the treatment of female infertility, as well as benign prostate hyperplasia represent attractive indications for Centrorelix. A depot-formulation for long-term treatment is under development.

Cetrorelix (Cetrotide®) was submitted to EU authorities for registration in controlled ovulation stimulation for article reproduction (COS/AT).

Results and discussion

The ongoing search for antagonists with further improved characteristics, such as increased duration of action, enhanced solubility and oral activity is stimulated by rapid progress in related sciences, like molecular biology. Now, the human pituitary receptor for Gonadorelin, a rhodopsin-like G-protein coupled heptahelical serpentine receptor, is available and at the beginning of a screening hierarchy.

Peptides can be synthesized and modified in a multitude of ways. Depending on the desired targets, physico-chemical properties may be finetuned by use of e.g. non-proteinogenic amino acids or molecular restrictions and by use of adequate formulations to prevent or manipulate biological interactions. In the case of pituitary-blocking LHRH peptides, the chemical nature of most powerful agonists as well as antagonists is somehow similar as a result of optimization procedures. Starting with the well characterized decapeptide antagonist Cetrorelix, different optimization strategies led to several new peptide lead structures with improved properties, such as enhanced affinity to the natural receptor, modest histamine release, extended in vivo-activity and favorable physicochemical properties (fig. 2).

Suitably protected peptide sequences were synthesized by standard solid phase peptide synthesis (SPPS) procedures, using Boc- in combination with Fmoc-strategy. Starting material was always 4-methylbenzhydrylamine (MBHA) resin, and cleavage of the final product was done with HF treatment (45 min. at 0 °C, scavenger). After HPLC purification, the compounds were isolated by lyophilization. The result of modifications by Deghenhgi led to a highly active decapeptide sequence with minimal histamine release and good water solubility. This structure, known as Teverlix (INN) (Antarelix™) [151277-78-5], differs from Cetrorelix in that it contains homocitrulline^6 instead of citrulline and isopropyllysine^8 instead of arginine^8.

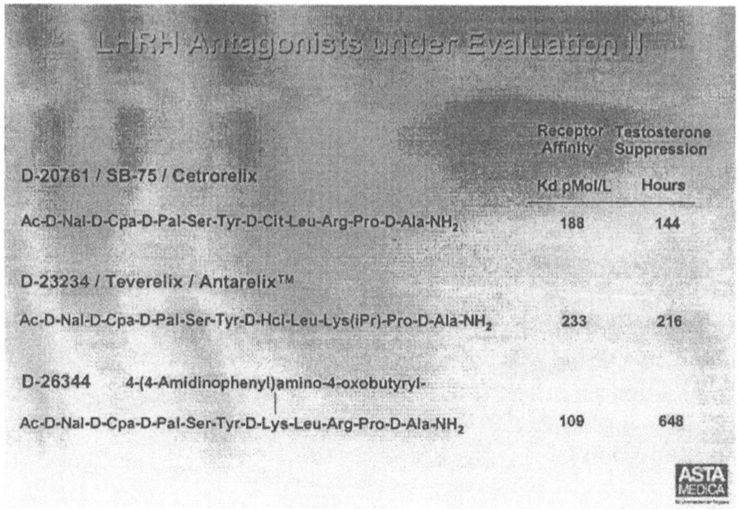

LHRH Antagonists under Evaluation II

	Receptor Affinity Kd pMol/L	Testosterone Suppression Hours
D-20761 / SB-75 / Cetrorelix		
Ac-D-Nal-D-Cpa-D-Pal-Ser-Tyr-D-Cit-Leu-Arg-Pro-D-Ala-NH_2	188	144
D-23234 / Teverelix / Antarelix™		
Ac-D-Nal-D-Cpa-D-Pal-Ser-Tyr-D-Hci-Leu-Lys(iPr)-Pro-D-Ala-NH_2	233	216
D-26344 4-(4-Amidinophenyl)amino-4-oxobutyryl-		
Ac-D-Nal-D-Cpa-D-Pal-Ser-Tyr-D-Lys-Leu-Arg-Pro-D-Ala-NH_2	109	648

ASTA MEDICA

Fig. 2.

The decapeptide analogue of Centrolix, D-26344, with a new D-Lys6-sidechain modification, shows human LHRH-receptor affinity in the range of Kd=109 pM (Cetrorelix: 188 pM), long lasting testosterone suppression in rats up to 648 hours (single doses, 1.5 mg/kg animal) and minimal histamine release.

Binding affinity of Teverrelix and D-26344 to the human LHRH receptor was determined in displacement binding experiments with 125I-Cetrorelix. The compounds had the best in vitro activity in comparison to other antagonists (fig. 2).

Lack of solubility and aggregation hampered the development of LHRH-antagonists in the past. D-26344 and Teverelix show minimal tendency for aggregation as measured by optical density in water over time (fig. 3). In addition, the stability of these antagonists against a series of enzymes was improved (fig. 4).

For some years, increased efforts have also been made to find substances with affinity to the LHRH receptor that do not have characteristic substance-specific properties and also the disadvantages of peptides (short half-life, lack of bioavailability), and yet have a high binding affinity. Ideally, such substances should be able to be administered orally, be sufficiently stable in the organism, and possess favorable pharmacological parameters comparable to peptide antagonists.

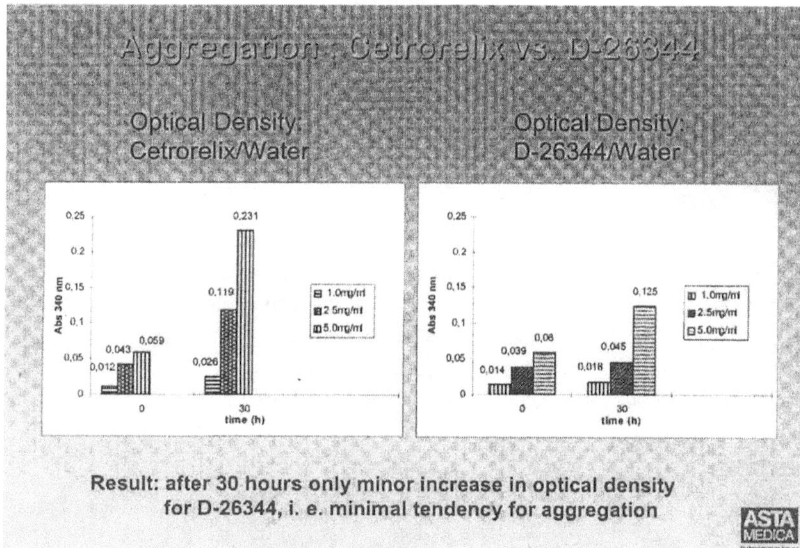
Aggregation : Cetrorelix vs. D-26344
Optical Density:
Cetrorelix/Water
Optical Density:
D-26344/Water
Abs 340 nm
0,25
0.2
0,15
0,1
0,05
0
0,012
0,043
0,059
0,026
0,119
0,231
0
30
time (h)
1.0mg/ml
2.5mg/ml
5.0mg/ml
0,014
0,039
0,06
0,018
0,045
0,125
Result: after 30 hours only minor increase in optical density
for D-26344, i. e. minimal tendency for aggregation
ASTA MEDICA

Fig. 3.

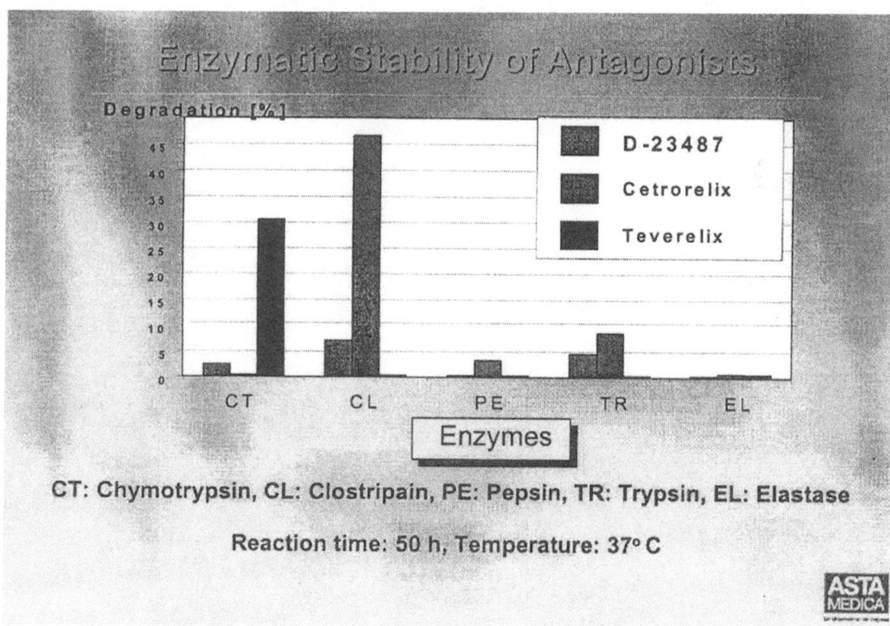
Enzymatic Stability of Antagonists
Degradation [%]
45
40
35
30
25
20
15
10
5
0
D-23487
Cetrorelix
Teverelix
CT
CL
PE
TR
EL
Enzymes
CT: Chymotrypsin, CL: Clostripain, PE: Pepsin, TR: Trypsin, EL: Elastase
Reaction time: 50 h, Temperature: 37° C
ASTA MEDICA

Fig. 4.

This strategy is confirmed by recent publications from the firm Takeda, who reported cyclic pentapeptides with affinity to the human LHRH receptor up to IC_{50} =0.07 μM. The search for lead structures is facilitated and will be further accelerated by use of compound libraries, high-throughput screening systems and better understanding of the 3D-structure of the human LHRH receptor.

Conclusion

Teverelix^{INN} (Antarelix) and D-26344, a D-Lys-6-analogue of Cetrorelix, show excellent human LHRH-receptor affinity, long lasting testosterone suppression in rats up to 648 hours (D-26344), and minimal histamine release. Both compounds are currently in preclinical evaluation at ASTA Medica for treatment of sex-hormone dependent tumors. Several non-malignant conditions such as benign prostatic hyperplasia are also possibly appreciate targets for these LHRH antagnist.

DNA-binding characteristics of antitumor drug actinomycin D and its analogs

Jian-Heng Shen, Xiao-Wu Yang, Zheng-Ping Jia and Rui Wang
Department of Biochemistry and Molecular Biology, School of Life Science, Lanzhou University, Lanzhou, 730000, China

Introduction

Actinomycin D (ActD, fig. 1), containing a planar phenoxazone ring and two cyclic pentapeptides, is one of the most intensely studied anticancer drug. It has been well known that the antitumor activity of ActD is due to its ability to bind to double-stranded DNA, resulting in the inhibition of DNA-directed RNA synthesis [1].

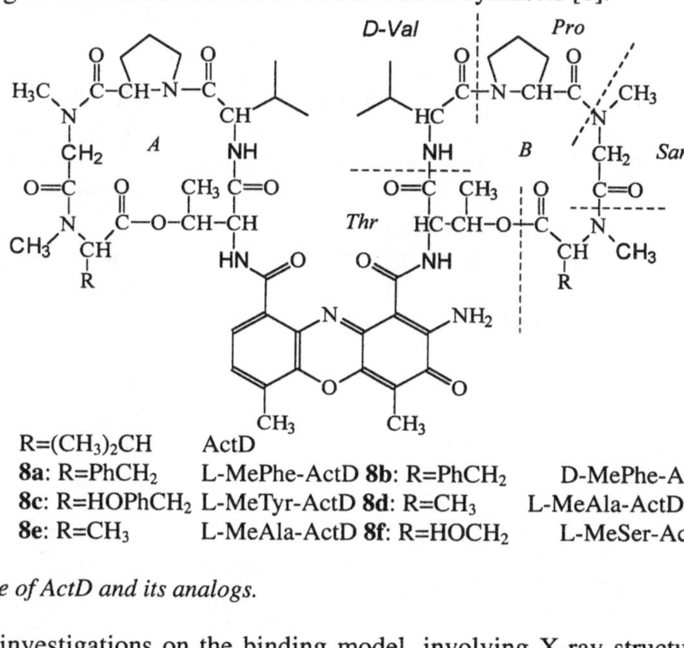

$R=(CH_3)_2CH$ ActD
8a: $R=PhCH_2$ L-MePhe-ActD **8b**: $R=PhCH_2$ D-MePhe-ActD
8c: $R=HOPhCH_2$ L-MeTyr-ActD **8d**: $R=CH_3$ L-MeAla-ActD
8e: $R=CH_3$ L-MeAla-ActD **8f**: $R=HOCH_2$ L-MeSer-ActD

Fig. 1. Structure of ActD and its analogs.

Numerous investigations on the binding model, involving X-ray structural analysis [2], 2D-NMR studies [3], and footprinting assays [4], have clearly shown that the phenoxazone ring intercalates into the sequence 5'-GC-3' from the minor groove of DNA, the two cyclic pentapeptide rings lie on both sides of the minor groove and cover four base pairs of DNA, and that the drug is tightly connected the DNA by forming four threonine-guanine hydrogen bonds and two additional hydrogen bonds between the N2 amino group of phenoxazone and the DNA backbone.

A few analogs modified N-methylvalines of the two pentapeptide rings have been shown very promising results. For example, MeLeu-ActD exhibits higher antitumor activity than ActD itself *in vitro* [5e]. Recent studies have also suggested that the N-methyl amino acids play an important role in antitumor activity [6] and the D-analog in which L-N-methylvalines was replaced by D-N-methylvalines showed higher DNA-binding affinity [7]. With this information in mind, We designed and totally synthesized some analogs of ActD replacing the N-methyl valine in the peptide moieties with N-methyl-L-(and D-)-phenylalanines, N-methyl-L(and D-)-alanines, N-methyl- tyrosine, and N-methyl-L-serine.The present study involves conformation and DNA-binding characteristics of these analogs. The antitumor activities of the ActD analogs *in vitro* have been examined using Human stomach Carcinoma AGC-7901 and hepatoma BEL-7402 cells; it was found that the antitumor activity was roughly correlated to those of DNA-binding characteristics. The results of acute toxicity showed that **8b, 8e** with methylvaline replaced with D-form amino acids have very 2~8 times bigger LD_{50} than the L-forms **8a, 8d** and ActD, and all analogs except for **8a, 8c** have lower toxicity than ActD itself.

Conformation

It is very interesting to examine the conformation change of the ActD analogs because the conformation often plays an important role on the biological activities. The conformations of ActD and its analogs were measured by CD spectra. As shown in Figure 2, all analogs exhibit the same negative cotton effect as well as ActD, suggesting that they have similar conformation [9]. There are slight difference between the aromatic analogs and the aliphatic analogs; the former show more strong absorption at 264 nm than the latter. There are no very obvious difference between the L-form analogs and the D-forms. The results can be supported by ^{1}HNMR spectra (experimental section) in which chemical shifts and NH, C2H coupling constants of the analogs are very closed to those of ActD, which has been variously interpreted by conformation analysis [10]. Upon the basis of these data, no significant conformational change resulting from the N-methyl amino acid replacements is apparent.

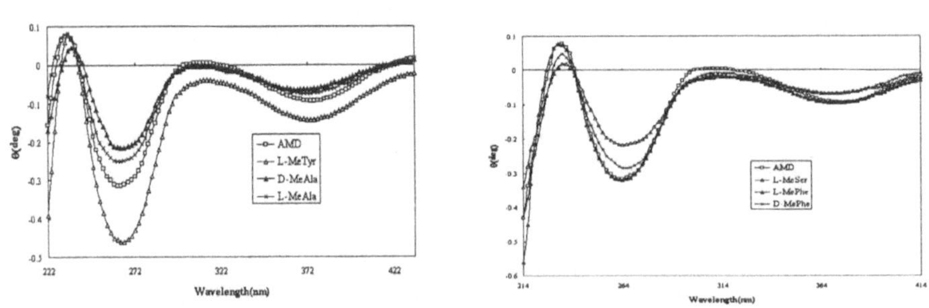

Fig. 2. CD spectra of ActD and its analogs.

DNA-Binding Characteristics

Studies with other ActD analogs has indicated that this measure of affinity to DNA correlates roughly with biological potency [11]; so, it is necessary to examine the DNA-binding characteristics. Different spectra of ActD analogs with calf thymus DNA are compared with that of ActD in Figure 3. All of ActD analogs show significant different spectra as ActD does, indicating that the analogs bind to the DNA [12]. However, the different spectra of the analogs are slightly different from each other. The aromatic analogs **8a~c** show the more significant spectra than the aliphatic **8d~f**, which suggests that **8a~c** bind to the DNA more strongly than the latter. Among them, L-MePhe-ActD **8a** exhibits the most significant spectra. There is no obvious difference between the L-form analogs and the D- forms.

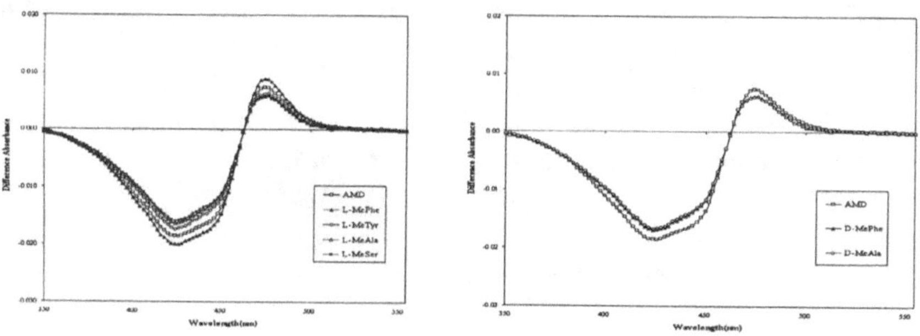

Fig. 3. Different spectra of ActD and its analogs.

Using the previous method [12], the association constants K's with intercalation mode, calculated from the visible spectra of ActD and its analogs titrated with Calf thymus DNA, were listed in Table 1. The association constant K, 1.3×10^5 of ActD with the DNA, is quite good agreement with the literature value, 1.3×10^5 [13b], suggesting that the method is acceptable. The K's value of the analogs is very clŏsed to that of ActD except for **8a**, which indicates that all of the analogs bind intercalately to the DNA. The analog **8a** binds to the DNA more strongly than ActD itself. The K's of the analog **8a~c** are slightly bigger than those of the aliphatic **8d~f**, in which the K's are nearly equivalent to each other. These results also support those of the different spectra. That is to say, the binding order of ActD and its analogs is: L-MePhe-ActD>ActD> L-MeTyr-ActD>D-MePhe-ActD>D-MeAla-ActD, L-MeAla-ActD and L-MeSer-ActD.

Table 1. Spectra analysis of ActD and its analogs titrated with Calf thymus DNA.

Run	Drugs	$K(10^5)$	H	$\Delta\lambda$(nm)	4	**8c**	1.25	0.40	22
1	**ActD**	1.30	0.4	19	5	**8d**	1.14	0.38	19
2	**8a**	1.38	0.44	22	6	**8e**	1.12	0.40	19
3	**8b**	1.20	0.39	20	7	**8f**	1.14	0.38	21

1) H is hypochromic effects of the spectra; 2) $\Delta\lambda$ □is red shifts of the spectra.

Experimental Section

CD and UV spectra measurements

All experiments were carried out at 15 °C with 1cm curette. CD spectra of ActD and its analogs in MeOH (6.5×10^{-5} μM) were obtained on Jasco-20 spectropolarimeter with absorbance change from 212 nm to 414 nm. UV/Visible spectra were measured on DW-2000 spectrometer. Calf thymus DNA was purchased from Shanghai factory of Biochemistry and purified to eliminate residual protein by two times phenol extraction and then ethanol precipitation ($\lambda_{260/280}$ >1.85). Stock solution for the DNA was prepared in phosphate buffer (0.02 M, pH 7.2), which was used for all binding experiments. The concentration of DNA stock solution was determined upon the basis of $\varepsilon_{260}=1.25\times10^4$ $M^{-1}cm^{-1}$ [13]. The solutions of drug-DNA complexes were keep at 37 °C for 30 min before measurement. The Visible spectra of the drugs titrated with the DNA were obtained by following procedure: After measurement of the absorbance of the drugs solution (2 μM), the solution of the drug-DNA complex was placed into sample cuvette and then the absorption spectrum was measured. This procedure was repeated six times so that six spectra were measured at different DNA concentrations (0, 10, 20, 30, 40, 50, and 60 μM) with the same drug concentration (2 μM). The absorbance was taken at 2nm intervals for calculation of the binding constants.

Acknowledgements

This work was supported by the National Natural Science Foundation of China.

References

1. a) Reich, E., Franklin, R.M., Shartkin, A.J. and Tatum, E.L., Science, 134 (1961) 536; b) Goldberg, I.H. and Rabinowitz, M., Science, 136 (1962) 315; c) Goldbrg, J.H. and Friedman, P.A., Annu. Rev. Biochem., 40 (1971) 775.
2. a) Takusagawa, F., Dabrow, M., Neidle, S. and Berman, H.M., Nature, 296 (1982) 466; b) Kamitori, S. and Takusagawa, F., J. Mol. Biol., 225 (1992) 445; c) Sobell, H.M. and Jain, S.C, J. Mol. Biol., 68 (1972) 21.
3. a) Brown, S.C., Mullis, K., Levenson, C. and Shafer, R.H., Biochemistry, 23 (1984) 403; b) Zhou, N., James, T.L. and Shafer, R.H., Biochemistry, 28 (1989) 5231; c) Liu, X., Chen, H. and Patel, D.J., J. Biomol. NMR., 1 (1991) 323; d) Chen, H., Liu, X. and Patel, D.J., J. Mol. Biol., 258 (1996) 457.
4. Goodisman, J., Rehfuss, R., Ward, B. and Dabrcwiak, J.C., Biochemistry, 31 (1992) 1046.
5. a) Meienbofer, J. and Atherton, E., In Perlman, D. (Eds.) Structure-activity Relationships among the semi-synthesis antibiotics, Academic Press, NewYork, 1977, p. 427; b) Sehgal, R. K., Almassian, B., Rasenbaum, D., Zadwzny, R. and Senguptes, S.K., J. Med. Chem., 31 (1988) 790; c) Wang, R., Chin. Chem. Lett., 6 (1995) 273; d) Mauger, A.B., Stuart. O.A., Ferretti, J.A. and Silverton, J.V., J. Am. Chem. Soc., 107 (1985) 7154; e) Mauger, A.B., Stuart, A., J. Med. Chem., 34 (1991) 1297.
6. Wang, R., Ni, J.M., Jia, Z.P., Pan, X.F. and Hu, X.Y., Chem. J. of Chin. Univ., 19 (1998)

7. Shinomyia, M., Chu, W., Carlson, R.G., Weavr, R.F. and Takusagawa, F., Biochemistry, 34 (1995) 8481.
8. When we finished the total synthesis of the analogs we find that the analogs of MePhe-ActD, MeTyr-ActD were synthesized. Biochemistry, 35 (1996) 13240.
9. a) Ascoli, F., Santis, D. and Savino, M., Nature, 227 (1970) 1237; b) Mosher, C.W. and Goodman, L., J. Org. Chem., 37 (1972) 2928.
10. a) Victor, T.A., Hruskn, F.E., Hikichi, K., Danyluk, S.S. and Bell, C.L., Nature, 223 (1969) 302; b) Conti, F. and Santis, P.D., Nature, 227 (1970) 1259; c) Lackner, H., Tetrahedron Lett., (1971) 4309; d) Lackner, H., Angew. Chem. Int. Ed. Engl., 14 (1975) 375.
11. a) Wang, R., Ni, J.M., Yang, X.W., Hu, X.Y. and Pan, X.F., In Xu, X.J., Ye, Y.H. and Tam, J.P.(Eds.) Peptides: Biology and Chemistry (Proceedings of the 1996 Chinese Peptide Symposium), Chengdu, China, 1996, p.53; b) Wadkins, R. M., and Jovin, T. M., Biochemistry, 30 (1991) 9469; c) Brennan, T.F. and Sengupta, S.K., J. Med. Chem., 26 (1983) 448; d) Sengupta, S. K., Anderson, J. E. and Kelley, C., J. Med. Chem., 25 (1982) 1214; e) Formica, J. V., Shatkin, A.J. and Katz, E., J. Bacteriol., 95 (1968) 2139.
12. a) Chu, W., Kamitori, S., Shinomiya, M., Carlson, R.C. and Takusagawa, F., J. Am. Chem. Soc. 116 (1994) 2243; b) Chu, W:, Shinomiya, M., Kamitori, K.Y., Kamitori, S., Carlson, R. G., Weavers, R.F. and Takusagawa, F., J. Am. Chem. Soc., 116 (1994) 7971.
13. Winkle, S.A. and Krugb, T.R., Nucleic Acids Res., 9 (1981) 3175.
14. Chen, F.M., Biochemistry, 27 (1988) 6393.

Design, synthesis of the peptide analogs of antitumor agent actinomycin D

Xiao-Wu Yang, Jian-Heng Shen, Zheng-Ping Jia and Rui Wang
Department of Biochemistry and Molecular Biology, School of Life Science, Lanzhou University, Lanzhou, 730000, China

Introduction

Actinomycin D (ActD, fig. 1), containing a planar phenoxazone ring and two cyclic pentapeptides, is one of the most intensely studied anticancer drugs and is currently used to treat highly malignant tumors such as Wilms'tumor and gestational choriocarcinoma [1]. Although ActD possesses high antitumor activities, its clinical usefulness is limited by its extreme cytotoxicity. Thus, if the structure of ActD can be modified to reduce its

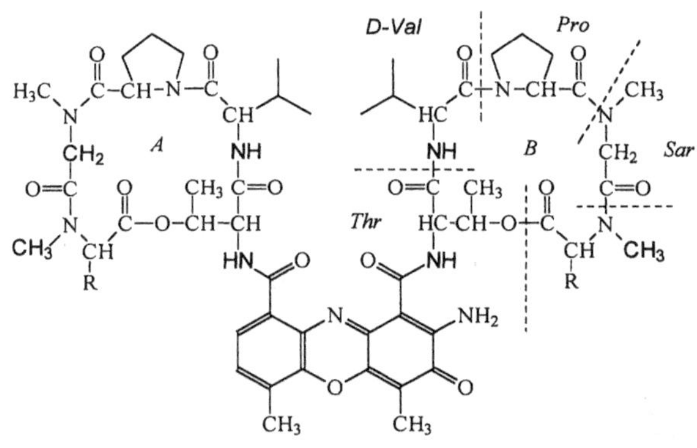

R=$(CH_3)_2CH$ ActD
8a: R=$PhCH_2$ L-MePhe-ActD **8b**: R=$PhCH_2$ D-MePhe-ActD
8c: R=$HOPhCH_2$ L-MeTyr-ActD **8d**: R=CH_3 L-MeAla-ActD
8e: R=CH_3 L-MeAla-ActD **8f**: R=$HOCH_2$ L-MeSer-ActD

Fig. 1. Structure of ActD and its analogs.

cytotoxicity while retaining its activity, such an analog would be a better antitumor drug. Analogs of ActD have been produced by directed biosynthesis, partial synthesis, and total synthesis [2]. In many cases, replacements of amino acid residues in the two cyclic pentapeptide rings have been found to render such analogs inactive or less active [2a]. We designed and totally synthesized some analogs of ActD replacing the N-methyl valine in the peptide moieties with N-methyl-L-(and D-)-phenylalanines, N-methyl-L(and D-)-alanines, N-methyl-tyrosine, and N-methyl-L-serine.

Results and Discussion

Synthesis

ActD analogs in which both N-methylvalines were replaced with L-MePhe, D-MePhe, L-MeAla, D-MeAla, MeTyr, and Me-Ser, have been totally synthesized [2c, 5] (Scheme1). All analogs are red solid and their structures have been identified with ^{1}HNMR,MS,and HRMS.

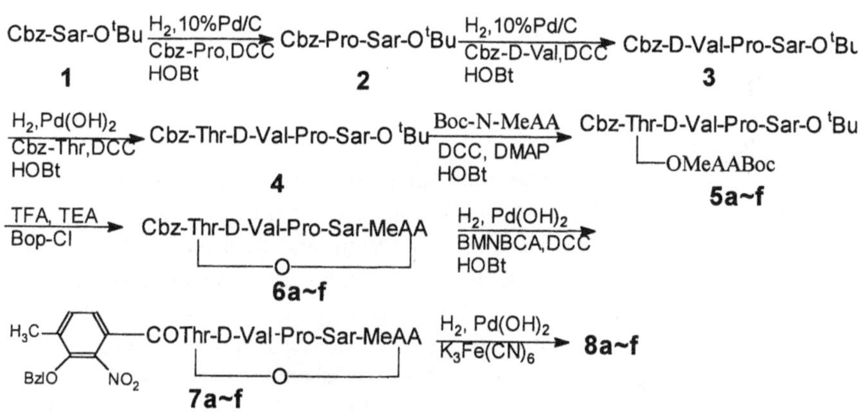

Scheme 1. Synthesis of ActD and its analogs.

Antitumor activity

The concentration of ActD and its analogs at 50% inhibition have been examined *in vitro* using Human stomach carinoma (SGC-7901) and heptoma BEL-7402 Cells, As listed in Table 2, all analogs except the analog **8f** show low IC_{50} at the same concentrations. The analog **8f** is almost inactive; the IC_{50} of the aromatic **8a~c** are very closed to ActD but are smaller than the aliphatic **8d~f**, which suggests that **8a~c** are more active. There is roughly correlation between DNA-binding characteristics and antitumer activities except for the L-form analogs **8a, 8d** which show stronger DNA-binding capacity but lower activities than the D-forms **8b, 8e**; however the difference is very small. This correlation is also reported by Mauger [6e]. It is necessary to note that there is very obvious change between the L-form analogs and the D-forms. The D-form analog **8b** with high activitiy but low toxicity (LD_{50} =2.56, which is 5 times lower than that of ActD). Though the LD_{50} of the D-form analog **8e** is bigger than the L-form **8d**, its activity is very low. The L-form and D-form analogs show similar CD spectra and different spectra but their LD_{50} is very different, suggesting that the configurtion of N-methyl amino acid residues of the cyclic peptide plays an important role on toxicity. The D-analog **8b** is very pressingly new antitumor drug.

Table 1. IC_{50} and LD_{50} of ActD and its analogs.

Run	Drug	IC_{50} (BEL-7402)	IC_{50} (SGC-7901)	LD_{50} (mg.kg-1)
1	ActD	1.31×10-7	7.22×10-8	0.51
2	**8a**	1.79×10-7	9.67×10-8	0.35
3	**8b**	1.43×10-7	5.08×10-8	2.56
4	**8c**	1.75×10-7	1.58×10-7	0.28
5	**8d**	3.31×10-7	2.85×10-7	2.30
6	**8e**	2.59×10-7	1.11×10-7	4.32
7	**8f**	7.33×10-7	1.12×10-6	8.77

Biological Evalution

Antitumor activitis *in vitro* were examined using Human stomach carcinoma SGC-7901 and hepatoma BEL-7402 Cells, which were diluted to an initial density of 1×10^5 mg/ml with RPMI 1640 medium with 15% fetal calf serum, and then placed in 96-well tissue culture plates and incibated at 37 °C with or without different dose of ActD and its analogs for 4 h. The OD values of MTT formazan produced by the viable cells were measured by a MR4000 spectrohotometer at 560 nm. The IC_{50} value were calculated by regression analysis. Acute toxicities of ActD and its analogs were test with F1 mice (BALB/c × LIBP/1) purchased from Lanzhou Institute of Biological Products. Six logarithmically spaced doses were injected *ip* into six groups of 6-week-old F1 mice (1:1 male/female, weighting 20 ± s 2 g). The LD_{50} values were calculated as previous method.

Experimental Section

Synthesis

All reactions were carried out under dry N_2 atmosphere unless otherwise stated and were measured by TLC with I_2 as developing agent. Anhydrous methene chloride (DCM) was distilled from calcium hydride. The products were isolated by flash column chromatograph (silica gel 60, particle size 20~40 um). Melting points were uncorrected; ^{1}HNMR spectra were obtained with Brucker 400 NMR spectrometer; FABMS spectra were recorded on a VG-ZAB mass spectrometer; Optical rotations were measured on WZZ automatic polarimeter. Synthesis of Linear teterapeptide refers to the literature 7. Abbreviations used are BMNBCA=3-(benzyloxyl)-4-methyl-2-nitrobenzoic acid, DCC = dicyclohexylcarbodiimide, DCM =methylene chloride, DMAP=4-(dimethyl amino) pyrridine, HOBt=1-hydroxybenzo triazole hydrate, PE=petroleum ether, TEA=triethyl amine, TFA=trifluroacetic acid, Z=benzoxycarbonyl.

General procedure for 5a-f

To a solution of linear tetrapeptide **4** (306 mg, 0.53 mM) and Boc-AA (0.78 mM) in dry DCM (4 mL) was added DMAP (49 mg, 0.4 mM) and then a solution of DCC (175 mg, 0.88 mM) in DCM (0.8 mL) at 0 °C. The mixture was stirred for 2 h at 0 °C and 20 h at 25 °C. After filtration, the filtrate was evaporated. The residue was dissolved in EtOAc (15 mL), and the solution was washed with 10% citric acid (2 × 5 mL), 10% $NaHCO_3$ (2 × 5 mL), and water (2 × 5 mL), and dried over anhydrous Na_2SO_4. After evaporation of the EtOAc, the pale yellow residue was prified by flash column chromatography on silica gel with EtOAc:petrolam ether to give the desired linear pentapeptide as white solid.

General procedure for 6a-f

A solution of linear **5** (0.41 mM) in TFA (3.8 mL) was stirred for 5 h at 0 °C. After removal of TFA *in vacuo*, the colourless residue was dissolved in EtOAc (40 mL) and then evaporated *in vacuo* The procedure was repeated three times to eliminate the residual TFA. Then, the residue was dissolved in dry DCM (40 mL) and the solution was added TEA (0.18 mL, 1.29 mM) at 0 °C. After stirring for 10 min, the mixture was diluted with DCM (160 mL) and then Bop-Cl (212 mg, 0.83 mM). The reaction mixture was stirred for 6'days at 20 °C, after evaporation of the solvent, the residue was dissolved in EtOAc (30 mL). The solution was washed with 10% citric acid (2 × 10 mL), 10% $NaHCO_3$ (2 × 10 mL), and water(2 × 10 mL), and dried over anhydrous Na_2SO_4. After remval of the EtOAc, the residue was purified by flash chromatography to obtained the cyclic pentapeptide as white solid.

General procedure for 7a-f

A solution of cyclic pentapeptide **6** (0.9 mM) in MeOH (15 mL) was hydrogenated over $Pd(OH)_2$ (30 mg) at 1atm for 1h with vigorous stirring. After filtration the filtrate was evaporated *in vacuo*. To a solution of the residue in DCM (8 mL) was added BMNBCA (96 mg, 0.33 mM) and HOBt (49 mg, 0.33 mM). The mixture was cooled to –10 °C for 5 h and kept at 20 °C for 10 h. After filtration, the solution was evaporated to give yellow residue which was dissolved in EtOAc (30 mL), washed with 10% citric acid (2 × 5 mL), 10%$NaHCO_3$ (3 × 5 mL), and water (5 mL), and dried over anhydrous Na_2SO_4. After evaporation of the EtOAc, the residue was prified by flash column chromatography with EtOAc methanel to obtain white powder.

General procedure for 8a-f

A solution of **7** (0.13 mM) in MeOH (10 mL) was hydrogenated over $Pd(OH)_2$ (15 mg) at room temperature for 3h. After filtration, the filtrate was added to a solution of phosphate buffer (28.5 mL, pH7.1) containing $K_3Fe(CN)_6$ (128 mg, 0.39 mM) with vigorous stirring . The mixture was stirred for 10 min at 25 °C following by adding water (59 mL). The reaction mixture was extracted with EtOAc (5 × 20 mL) and the extract was dried over anhydrous Na_2SO_4. after evaporation of the EtOAc, the residue was purified by flash column chromatography with EtOAc to give the red solid.

Acknowledgements

This work was supported by the National Natural Science Foundation of China.

References

1. a) Farber, S.J., J. Am. Med. Ass., 198 (1966) 826; b) Lewis, J.L., Cancer, 30 (1972) 1517; c) Marina, W., Fontanesi, J., Kun, L., Rao, B., Tenkins, J.J., Thompson, E.I. and Etcubanas, E., Cancer, 70 (1992) 2568.
2. a) Meienbofer, J., Atherton, E. and Perlman, D., Stracture-activity Relationships among the semisynthesis antibiotics. Academic Press, NewYork, 1977, p. 427; b) Sehgal, R. K., Almassian, B., Rasenbaum, D., Zadwzny, R. and Senguptes, S.K., J. Med. Chem., 31 (1988) 790; c) Wang, R., Chin. Chem. Lett., 6 (1995) 273; d) Mauger, A. B., Stuart. O.A., Ferretti, J.A. and Silverton, J.V., J. Am. Chem. Soc., 107 (1985) 7154; e) Mauger, A.B. and Stuart, A., J. Med. Chem., 34 (1991) 1297.
3. Wang, R., Ni, J.M., Jia, Z.P., Pan, X.F. and Hu, X.Y., Chem. J. of Chin. Univ., 19 (1998) 243.
4. Shinomyia, M., Chu, W., Carlson, R.G., Weavr, R.F. and Takusagawa, F., Biochemistry, 34 (1995) 8481.
5. When we finished the total synthesis of the analogs we find that the analogs of MePhe-ActD, MeTyr-ActD were synthesized. Biochemistry, 35 (1996) 13240.
6. a) Ascoli, F., Santis, D. and Savino, M., Nature, 227 (1970) 1237; b) Mosher, C.W. and Goodman, L., J. Org. Chem., 37 (1972) 2928
7. a) Victor, T.A., Hruskn, F.E., Hikichi, K., Danyluk, S.S. and Bell, C.L., Nature, 223 (1969) 302; b) Conti, F. and Santis, P.D., Nature, 227 (1970) 1259; c) Lackner, H., Tetrahedron Lett., (1971) 4309; d) Lackner, H., Engew. Chem. Int. Ed. Engl., 14 (1975) 375.
8. a) Wang, R., Ni, J.M., Yang, X.W., Hu, X.Y. and Pan, X.F., In Xu, X. J., Ye, Y. H. and Tam, J.P. (Eds.) Peptides: Biology and Chemistry (Proceedings of the 1996 Chinese Peptide Symposium), Chengdu, China, 1996,p.53; b) Wadkins, R.M. and Jovin, T.M., Biochemistry, 30 (1991) 9469; c) Brennan, T.F. and Sengupta, S.K., J. Med. Chem., 26 (1983) 448; d) Sengupta, S.K., Anderson, J.E. and Kelley, C., J. Med. Chem., 25 (1982) 1214; e) Formica, J. V., Shatkin, A.J. and Katz, E., J. Bacteriol., 95 (1968) 2139.
9. a) Chu, W., Kamitori, S., Shinomiya, M., Carlson, R.C. and Takusagawa, F., J. Am. Chem. Soc., 116 (1994) 2243; b) Chu, W., Shinomiya, M., Kamitori, K.Y., Kamitori, S., Carlson, R.G., Weavers, R.F. and Takusagawa, F., J. Am. Chem. Soc., 116 (1994) 7971.
10. Winkle, S.A. and Krugb, T.R., Nucleic Acids Res., 9 (1981) 3175.
11. Chen, F.M., Biochemistry, 27 (1988) 6393.

Authors Index

Subject Index

SPRINGER NATURE

GPSR Compliance

The European Union's (EU) General Product Safety Regulation (GPSR) is a set of rules that requires consumer products to be safe and our obligations to ensure this.

If you have any concerns about our products, you can contact us on ProductSafety@springernature.com

In case Publisher is established outside the EU, the EU authorized representative is:

Springer Nature Customer Service Center GmbH
Europaplatz 3
69115 Heidelberg, Germany

Kolibri 360 GmbH
Ferdinand-Jühlke-Straße 7
99095 Erfurt, Deutschland
produktsicherheit@kolibri360.de